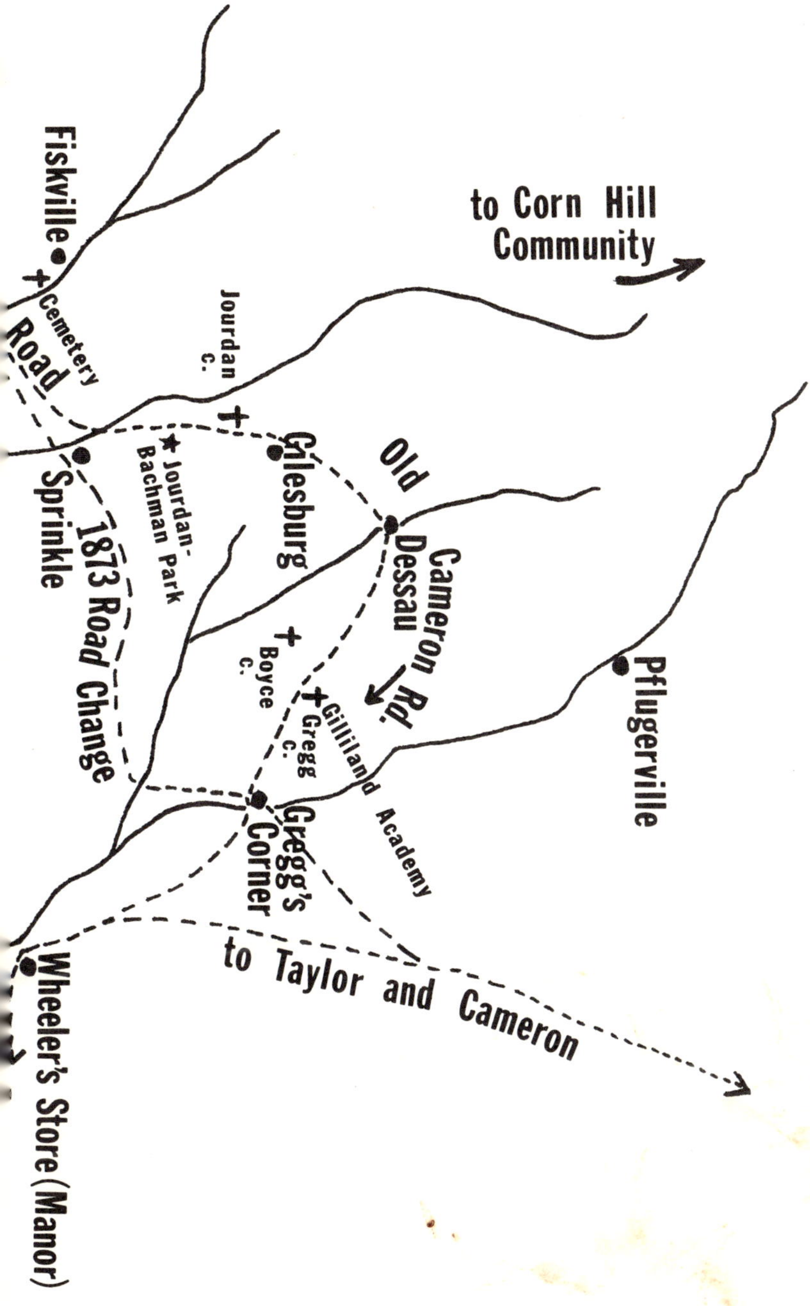
to Corn Hill Community
Fiskville
Cemetery
Road
Jourdan C.
Jourdan-Bachman Park
Gilesburg
Sprinkle
1873 Road Change
Old Dessau
Cameron Rd.
Boyce C.
Gilliland Academy
Gregg C.
Pflugerville
Gregg's Corner
Wheeler's Store (Manor)
to Taylor and Cameron

Kingdom Come!

Kingdom Go!

The Great Spirit

M. Jourdan Atkinson
and
Eugene V. Giles

EAKIN PRESS ★ BURNET, TEXAS

Published in the United States of America
By Eakin Press, P.O. Box 178, Burnet, Texas 78611

ISBN 0-89015-243-8

FIRST EDITION

ACKNOWLEDGMENTS

To all whose names appear as references in *Kingdom* I am deeply indebted, especially so to my first cousin-one-removed, Pauline Payton Johnson, and to Onie Jackson Kruger—a cousin of sorts (her grandfather Sprinkle was my father's sister Mary's second husband, she his second wife) who enabled me to recall a childhood long lost to time and distance. There was Preshy Barton, a cousin on Harriet's side, who searched Travis County Deed Records; Pat F. Cochran, a cousin on Frederic's side, who read the *ms.* to check data; the librarians: Mrs. Frances Moore in the Austin and Travis County Collection, and Mrs. Loraine Jeter in the Houston Public Library's Texas Room; Loraine B. Webb of the Houston Medical Center who typed the *ms.* twice; and the several proof readers: Dr. Christine V. Brannan, Mrs. Leonia L. Dorsey, and my granddaughters, Doris Wise who read the first part, and Lee Wise, who, like the darling she is (and the linguist), gave a week from her Christmas vacation as a college senior to the final reading. Lastly, there was Andrew Forest Muir whose expertise I shall always miss in building a Texas history background. Then, where everything else was done, Jennifer Bird, through familiarity with the Southwest, lent the imagery of her drawings to a past-and-gone era between Walnut and Gilliland Creeks, the area where she and John Bird have recently built a home on Gregg Lane.

M. Jourdan Atkinson

Pale ink is better than the best memory.

—Confucious

Too much somber fact is hard on glamour, especially the post-mortem and legendary kind.

—George Woodbury,
John Goffe's Legacy

The Papers and The Book

If, in the Spring, a young man's fancy turns to romance, so, in the autumn, an old man's thought dwells on memorabilia.
—M. Atkinson

The Giles Papers are a medley of store accounts, farm accounts, dairy accounts, letters, diaries, legislative house bills and personal essays. The nucleus is a set of fourteen ledgers, the first two for 1899 and 1901. These two are account books, strictly. The succeeding twelve, many of them dated for more than one year, run until 1952.

In all except the two earliest ledgers my cousin Eugene went back, as age came on, to fill in blank pages or parts of pages with his Memoirs, writing sometimes from front to back, sometimes from back to front, at other times just skipping about wherever he found a space. Unfortunately, before being persuaded of a possible historical value, he burned a part of the family correspondence; a pity, for it had been family habit to preserve all documentation. This partial burning accounts, perhaps, for there being much from him and from Laura but almost nothing from Alice.

His writing style is always eloquent; however, he was a speaker rather than a writer, and in the Memoirs which contain some of his most interesting observations, he is handicapped by an inability of pen or pencil to race with his thoughts. Except where he rewrote some of the essays in old tablets, he did not pause to order tenses and usually found a dangling participle the quickest way to elide a sentence (thus forcing the reader to backtrack for its meaning). But, after all, at such times he was only thinking of himself—savoring the past. When he did come to appreciate what he had, he left the Papers to me with a sum of money to shape the content into form. The result is *Kingdom Come! Kingdom Go!*

A generation gap between us accounts for our differing points of view in the historical perspective while a mutual familiarity with the Walnut and Gilliland Creeks locale accounts for similarities in our outlooks. The basic Papers will be added to the Austin and Travis County Collection in Eugene's name. In the book, where factual material was lacking to maintain a running story of the life and times, it has been necessary to fill in imaginatively; however, such detail has been held to a minimum keeping with the Indian pledge:

I tell the tale as 'twas told to me,
The factual with the fantasy;
May the truths I have told go bold and free
And the lies I have told come home to me.

M. Jourdan Atkinson

FREDERIC AND HARRIET'S house on Walnut after it became "the Old Place."

FREDERIC.

HARRIET.

To

Frederic and Harriet,

The Pioneers

who, had they been Europeans instead of American Europoids, might have founded a dynasty. As it was, they founded a family that extends from Canada to the Gulf of Mexico and from ocean to ocean. May their spirits be pleased with this book, the joint work of their last grandson, Eugene Victor Giles (pronounced like "j" in "John" and not like "g" in "Gideon"), who died in 1974, and their last granddaughter, a generation younger than Eugene—and the only grandchild who never knew them.

Apologia
To the Reader

who will find these chapters following no format established for biographies, genealogies or scientific treaties but skipping, more in the order of a series of paintings, from one vignette to another—the effort (with such documentation as is given in the footnotes) being to portray, along with the extension of a family, the rise and fall of a cotton kingdom in Central Texas—a period of a hundred years.

CONTENTS

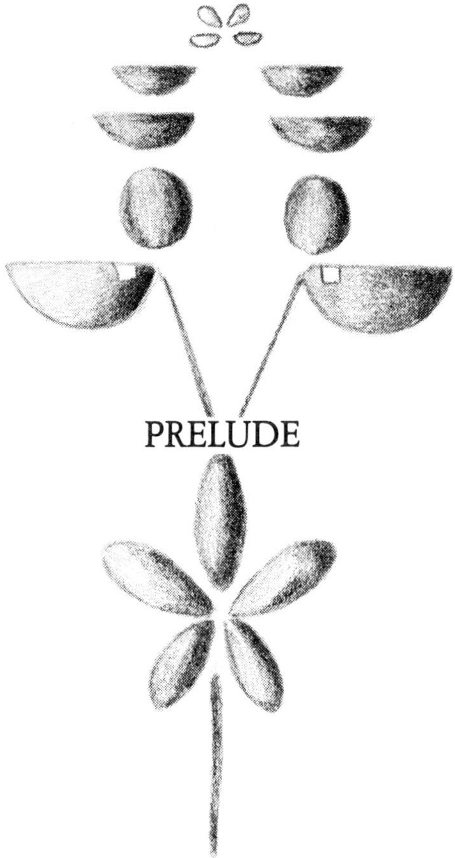

PRELUDE

Like the Castilian insignia among our state seals, like the Moorish patterns elaborating our boots, like the liquid words embroidering our Texanease patois (lingo), the blue lupine birthmarks our Spanish heritage.

The Wheels of Progress

The beat of horses' hooves came first, where there had been no horses—only soundless mocassins. Then the rumbling of carts meshed with the multiple tread of oxen. For a hundred years cart wheels would be the only wheels rolling across Texas, moving Franciscan friars and their *vecinos* east. Then would come the wheels of the great wagons (prairie schooners) moving other settlers west.

*Las Carretas de Rueda de Panoche**

Iglehart, in *A Boy Captive of the Texas Mier Expedition,* describes one of these "towering, squeaky, pondorous carts" with "pancake" wheels—the wheels made of three pieces of timber mortised together, the tires of rawhide—as coming "piled with corn and golden pumpkins, groaning and creaking, filling the air with discordant squeaks and squeals." (A different writer notes a use of prickly pear pads in lieu of axle grease to cut down on the racket and relieve the wheels' overstrain.) C.L. Bock drew for this book (San Antonio, 1909) an empty cart, side-boarded with rawhide fast to poles rising all round from the cart's bottom to hold in a cargo. Such a cart, fully loaded, would require six or eight oxen, yokes bound to their horns (instead of fitted to their necks, Anglo-fashion) to keep the raucous wheels in motion. Such was, and had been for generations, the transport system of Mexico and Texas until the Cart War of 1857 when the Great American Bull Whacker—the Anglo teamster—descended in force, with Anglo-Saxon expletive and a mighty cracking of bull whip, to banish native drivers from the land above the Rio Grande.

The Prairie Schooners

> There were from two to four yoke of oxen to each wagon. The wagons were covered with white sheeting drawn over bows, standing high, sheet drawn by a cord, until only a hole was left for a lookout—open front. It looked like a veritable telescope and a telescope it was, magnifying everything before and minifying everything behind.
>
> —G. W. Kennedy, *The Pioneer Campfire*

**Rueda de panoche:* pancake wheel.

The Black Land and The Spanish Heritage

It was in Texas . . . that the most extensive and intimate associations between the English-speaking and the Spanish-speaking peoples of this continent took place . . . Linguistic borrowings by English in the lone star state have been more numerous than in any other section of the country.

—Harold Woodmansee Bentley,
A Dictionary of Spanish Terms in English

It was a fair land, the black prairies rolling below the white overhang of limestone brows the Spanish called *balcones.* It was blanketed with wild-flowers in April and May—green the rest of the time, if it rained enough, or golden with sunburnt grasses the Indians called Mother Earth's hair—except when freezing northers blew their brief blasts. It was the winter home of buffalo and the year-round home of myriad deer.

It was a narrow strip between limestone ledges to westward and a low *cuesta* (bench or ridge) to eastward. The Comanches, peering down from the *balcones* (which were theirs) wanted it. The Apaches, raiding in from the farther-off western aridity, wanted it. The Tonkawa had it and, where artesian springs spouted from under the limestone brows, had beaded it with towns the Spanish called *rancherías.*

The Spanish spent a century monumentalizing those towns with missions. The trail they laid, marked in spring by blue lupine, stretched from Mexican mines south of the Rio Grande to French forts fringing out from the Mississippi. Laboring oxen seeded the flower in their tracks with droppings from hay the Crown shipped over from Spain to maintain a traffic asserting its hold on a territory the French would have liked to claim by virtue of lost La Salle and the French-Indian trade. Europe knew Texas, the land between, as the Pyrennes of the Americas.

The *vecinos* (vay-seé-nos: Spanish settlers) called the flower *el conejo* (co-náy-ho) for the cottontail rabbit, because of its white tip.

Botanists who knew its kind as the Wolf Flower, because it appeared to devour the soil, turning the land into a mirror image of the sky, would identify *el conejo* as *Lupinus texensis.*[1] The first Anglos to come called it "buffalo clover" for the humped herds they found dotting the blue pastures below the *balcones;* their descendants would rename it "the bluebonnet" for their pioneer ancestress' headgear.

The *vecinos* and the Franciscan friars were gone when these early Anglos came, thanks to the Spanish Crown's treatment of its *criollas*[2] (American-born subjects); but the Tonkawa were there, at war with the Comanchería and with the Apachería, as they had been before the Spanish came. Except for the blue flowers, the prairies were as they had been, ribboned with creeks flowing into rivers that flowed to the sea. Two creeks in particular caught the Anglo eye and were given Anglo names: Gilliland, for the first pioneer family to settle thereon, and Walnut, for the grape-hung groves of black walnut and pecan along its banks.

The cleft of the Balcones Escarpment bisects Texas rainbow-wise, southwest to northeast, from the Rio Grande to the Red River. Along its major portion lies the narrow black strip that, when the Civil War began, was just being riven for cotton. When World War I came, these fields were being ravished by alien Johnson grass that might more justly have been called lupine or "wolf" than the innocent blue *conejo.*

Even so, when World War II began, ninety percent of Texas' population (according to demographic charts made by the state's planning board) had gathered onto the black prairies that were but recently, and all too briefly, a mighty province in the realm of King Cotton.

The older towns are now cities—for man will eternally build, as did the Tonkawa and the Spanish, where water is.[3]

---

1. Schulz, *Texas Wild Flowers.*

2. *Criolla, criol',* from *criar de los colonias* (to be born in the colonies). "Creole," French colonial adaptation from the Spanish.

3. "The Spanish": Spanish-speaking ethnics native to the country *(criol'* or *mestizo)* as opposed to Spaniards (also called Peninsularies) direct from overseas. *Mestizos:* of mixed Spanish and Indian heritage.

PART ONE

The Texicans

They were "Coahuiltexians" when the Mexican states of Coahuila-and-Texas were Siamese twins, Coahuila reaching almost to San Antonio and Saltillo as their joint capital.* After the disjoining in 1836, "Texican" (not "Tejano" or "Texian"—"Texian" came later) was how Anglos above the Rio Grande distinguished themselves. There may have been a confusion in the minds of Easterners where Texas was referred to as "the Bantam Republic"—and where A Comprehensive Atlas, published in New York and Boston, uses "Texasians," but in family letters written during the Civil War, the use is "Texican."*

—M. Atkinson

*Gammel's *Laws of Texas,* Vol. I, page 64.

Frederic and Harriet

They said Texas was "held together with rawhide."

By the last day of the year 1839, Frederic and Harriet Jourdan had reached the home of John and Maria Atkinson in Bell's Settlement on Mill Creek, Austin County, in the depth of the Republic of Texas. Besides their small children: Harriet Emily, George Washington and Mary Catherine, they had with them Harriet and Maria's father, Old John Bachman, and his two young sons, John Beneke and William Gordon.[1] They came none too soon. December 31st was the final day the Board of Land commissioners at *San Felipe de Austin* (the county seat) would issue conditional certificates for third-class head-rights of 640 acres on the northwesterly frontier—the conditions to be removed after three years' Texas residence.[2] Maria, Old John's eldest, had been

---

1. The sisters were born in South Carolina. Their mother, Mary Catherine Barbara Beneke, deceased when Harriet was twelve, was from Barnwell County; their father, John Bachman, from Orange County.

Data provided by P. F. Cochran, Houston, shows that on June 26, 1838, Harriet released her dower right to 80 acres of land, Carroll County, Mississippi, to one Hardy Sanders (Carroll County Records, Deed Book B, p. 165). On November 7, 1838, (same book, p. 571) her father made over his personal properties to his thirty-eight-year-old son, Charles, stipulating that Charles see to the education of sixteen-year-old William Gordon and look after their stepmother, who, previous to her second marriage, had been Mrs. Rachel Carraway. This was apparently a proviso for an extended absence, even a possible death, inasmuch as Gordon was in Texas the next year. Charles remanded the properties to his father on February 26, 1846 (Deed Book G, p. 111) following Old John's return with the boys in Mississippi.

From these documents and later acknowledgements to William Gordon and John Beneke from the Republic of Texas, it is assumed by P.F. Cochran, great-grandson of Gordon, and also Minnie Ray Swofford, Lubbock, Texas, granddaughter of John Beneke, that both young men and their father came

fourteen when Harriet, eighth of his eleven offspring (John Beneke and William Gordon were tenth and eleventh) was born in South Carolina in 1815.

The Atkinsons had arrived on February 14th, 1835. Texas was not then disjoined from the barely independent Republic of Mexico, but would be the next year, when *el presidente,* Don Antonio López de Santa Anna, in fear for his life, ceded the country beyond the Rio Grande to a handful of Anglo and Spanish colonials on the Battlefield of San Jacinto. The Atkinsons brought with them five daughters and three sons; their first-born, William, and a baby, Martha Ann, were left in graves beyond the Mississippi. A seventh daughter was born on October 14, 1836. Another child was expected in January of '40.[3]

Due to their whylom colonial status, the Republic had issued the Atkinsons an unconditional headright to a league and a *labór* of land in Gonzales County and John would receive title to 360 acres in the Settlement when bounties for military service were paid off.[4] Dubbed "The Bantam Republic" by eastern newspapers and periodicals, Texas had nothing with which to pay its public debt other than millions and millions of acres the Pope, endowed by God, had bestowed on the Spanish Crown in 1492; the Crown had lost to a rebelling Mexico in 1821 and Santa Anna had ceded away while under duress in 1836.

Present through what was certain to become history, the Atkinson children chattered of the part they and their mother had played in what they called the Runaway Scrape while their father was on the way back from Alabama with supplies for the colonial army.[5] As they and

---

out to Texas for a visit with the Atkinsons when Frederic and Harriet came out for permanent residence.

2. According to land office records for the Republic of Texas, 1st class headrights (unconditional) were issued to citizens whose residence preceded the signing of the Declaration of Independence on March 2nd, 1836, and were for a league and a labór of land (approximately 5000 acres). Conditional headrights, 2nd class, for 1280 acres, were issued to families arriving before January 1st, 1841. Single men received half as much as married men. All conditional certificates were canceled after three years' residence, allowing for permanent titles. (Old Spanish Crown grants for *haciendas:* ah-see-én-dahs—"h" is silent in Spanish—covered 30,000 acres.)

3. Family data from John Brooks, civil engineer, Bellville, Texas, who was John and Maria's grandson.

4. A league of land, as rated by the Texas land office, was 4428.4 acres; a labór, 177.1 acres. In Spanish terms, labórs were tillable land, the rest pasturage, or "wild land."

5. A question exists as to John Atkinson's whereabouts at this time. Since

everyone else told the story, it had been a great adventure:

> They were startled as they pursued their everyday affairs to have General Sam Houston's courier come dashing into the Settlement ordering immediate evacuation! Shouting that the Alamo had fallen Sunday! (March 6) That Houston had the army on the march to the Brazos! That Santa Anna was on the way to loot and burn! That Comanches were massed on the Hill Line, waiting to come down! That when the armies crossed the river only Tonks would be left to protect the prairies! With such a hullabaloo! in their ears there was nothing to do but go—and nothing to go with but the water slide and one yoke of oxen—their father had the wagon.[6]
>
> Sixteen-year-old Sarah, fourteen-year-old Elizabeth and ten-year-old Catherine Virginia, with John Pinkney, thirteen and Alexander, seven, tipped the water barrel off the slide, brought the oxen up and yoked them to it while their mother buried her pots and pans in a hastily scraped out garden plot and scattered mustard seed over the disturbed earth. On the slide went the little ones: five-year-old Nancy Durany, three-year-old Martha and Henry Turner, a year and nine months, together with the precious Seth Thomas clock and some balls of spun cotton—everything that could be loaded and still leave room for a scant bundle of food and clothing.[7]
>
> Almost within minutes they were headed for the river and Groce's Ferry fifteen miles away over the prairie. Here General

---

his name is on the San Jacinto Monument Honor Roll, listed with Captain Gibson Kuykeydall's company, which was detailed to remain in charge of baggage and supplies at Harrisburg because many of the company were down of the measles when the rest of the army rowed across to the peninsula where Houston staged his coup, family assumption has been that he was with the Measaleers, but the Brooks ms. says otherwise. Hence it is here assumed that he had gone to the States for supplies (see Note 3 above). The ms. tallies with the writer's memory of asking, as a child, "Where were our folks at the time of San Jacinto?" and being told, "They were on their way here—they joined the army that was seeing the Mexicans out of Texas as soon as they got here." Use of this bit of family lore in a hastily composed chapter to meet a deadline in *Women of Early Texas* exemplifies the sin of going to publication with insufficient supporting research—for the writer took this to mean *Frederic* joined the army.

6. Waterslide: a crude sled made of heavy timbers and mounted with a large water barrel for hauling up drinking water. Sometimes merely a forked log (called "a water lizard") had to serve.

7. Maria's eleventh child, Louisa Jane, was born in October following the Runaway Scrape. She would have three others: Rebecca Caroline, Solomon and James. Both Rebecca Caroline and James died in infancy.

Houston would drill his make-shift army, perhaps six hundred in number—it was said with never more than a third present for any roll call, one third being down sick and another third riding off at will to see about their families. Jared Groce, the colony's greatest planter, though on a sickbed and though he had had no part in the quarrel that had led the state of Zacatecas and the Siamese-sister states of Texas-and-Coahuila to rebellion, would come from his "Retreat" in Grimes County, whither he had removed for his health, to help his son, Leonard, provision Houston's men and horses.[8] The latter advised Maria to hasten east in the wake of Coahuila's refugee government, then established at Nacogdoches in the thick of the piney woods, but prepared to move on into Louisiana should Santa Anna cross the landmark Trinity River.[9] Saying she could not hope to escape Comanches with a water slide, Mr. Groce lent her a high-wheeled Mexican cart and stored the cotton yarn against her return.[10]

It was a momentous flight through spring rains. Striking the main road east (a trail, really) the family became part of an exodus of women and children who poured into the road from all directions—many on foot driving what stock they had been able to muster, with dogs running in and out among the muddy hooves. Some, the more fortunate, rolled up, like the Atkinsons, on oxcart or in covered wagons; some had Blacks along. Here and there was a child mounted on a milk cow. There were no horses—the horses were with the army. At one trail crossing a woman was heard from afar shouting to her ox. The single ox, Blossom, dragged a rawhide to join the throng. Couched on the rawhide were a tiny boy and girl snuggled in a blanket against a side of bacon and a sack of corn meal. The hide of Blossom's mate, Bob Tail, had been stretched, drying, on the cabin wall

8. Jared Groce died November 30th, 1836. Items here relating to him are from Sarah Wharton Groce Berlet's ms.: "Jared Ellison Groce, One of the Old Three Hundred." (The "Old Three Hundred": earliest Anglo colony to form in Texas.)

9. *Coahuila:* có-a-weé-lah. The revolution began a year before San Jacinto (sán jay-sín-tow, as Texas pronounces it) with a revolt of the state of Zacatecas against Santa Anna's self-imposed military dictatorship. The revolt was joined by the Siamese-sister states, Texas-and-Coahuila, under their flag—twin stars on a blue field. Part of their grievance was removal of their joint capital from Monclova to Saltillo. Texas was, at the time, divided into three departments, with subcapitals at *San Antonio de Bexar* (pronounced báy-er, as in aspirin); Nacogdoches (náck-og-dóugh-chess) for the department of that name, which was the oldest of the three; and at San Felipe de Austin, for the Department of the Brazos.

10. Brooks ms.

when Houston's courier galloped by. To snatch it down, hack a hole in the neck and run a rope through the hole to Blossom's yoke was the only hope of escape.[11]

By the third week of April, the trek was well into the pines. There was sickness after more than a month of exposure: shivering around campfires when the wind blew, sleeping damp, eating what could be got together and cooked at night—but behind were the armies, and behind the armies, the ashes of cabins. It was in this frame of mind that the straggling *cavayard* was overtaken by Houston's courier, this time shouting, unbelievably, "Turn back! Turn back! The Mexicans are leaving Texas!"[12]

---

11. The story of Mrs. Blocker and her rawhide was told to the writer by a classmate at The University of Texas, Viola Fontaine Corley, of Mexia (may-hǎy-ah), Texas, who had it from her mother, who had been Viola Fontaine. It has not been possible to identify Mrs. Blocker further, despite considerable effort.

12. *Cavayard:* a Texanism derived from the Spanish *caballada* (cŏb-al-yǎh-thah). It is found in old narrations; the writer's father used it often to denote horses or people strung out along a road, much as one might say "a pascel" of one or the other.

Although Santa Anna was accused of selling Mexican Indians into South American slavery by the boatload at twenty-five pesos a head, he entered Texas with the avowed purpose of freeing all slaves, in accord with the prevailing constitution. A *fuero* (fwǎy-rŏw-legal exemption) had allowed Stephen F. Austin's colonists to bring their Blacks into the Department of the Brazos. This jumble of legalities (and *el presidente's* policy of the moment) made any Black or near-Black a logical go-between-the-armies for the period; hence they were used by both sides as couriers. The victory this courier was announcing (qualified by some historians as the world's sixteenth decisive battle) had been over in sixteen minutes. Santa Anna's contingent of seven hundred to a thousand men, augmented by his brother-in-law, General Cos, who led in an additional four hundred (following an all-night march), attacked during the siesta hour, scattered helter-skelter to be trampled into the marshes under pounding hooves of Texican horses; with the exception, that is, of Colonel Juan Almonte's brigade.

Almonte, Santa Anna's *aide-de-camp* (son of the priest-general, Morelos, who sent him to the States to be educated) had advised against taking the central contingent from the main army and leaving the semi-arid country around San Antonio since his commander-in-chief had no experience in fighting in mud and muck, but his advice was waved aside with the boast that, "I will drive these settlers before me like smoke before the wind!" Consequently, Almonte rode his charger back and forth in front of his tents during the debacle, demanding to surrender his men "after the fashion of civilized warfare," thus gaining the everlasting respect of cursing Texans (known thereafter

Maria, once more at Groce's Crossing, found that Houston's men, in their need, had taken the weights from her clock to mold into bullets, her cotton yarn to braid into hackamores (bitless bridles or *bosals*) for their horses.[13] But what matter—hadn't the war been won in sixteen minutes? Wouldn't the clock tick again if the weights were restored?[14] And Santa Anna had missed the Settlement—her cabin was still standing, her pots and pans safe under their greening coverlid.

It was a long story. It would take days to tell it all, what with one remembering this and another that; meanwhile, 1840 brought, besides a new Atkinson baby, Rebecca Caroline, a new subject for discussion at the Atkinson fireside.

John Beneke, in his twenty-first year and fresh to adventure, volunteered to take John Atkinson's place in Captain John Bird's company of Texas Rangers, about to depart for the western frontier. The company was hardly encamped when, to his chagrin, an accidental gunshot wound sent him home with a temporarily useless hand and a

---

by terrified women cooks as "*estos señores* God Damns!"). Meanwhile, *el presidente* (who probably never knew the enemy handful had been increased to near eight hundred by some U. S. regulars who, though forbidden to take part, came across from the Louisiana-Texas border) crawled off through sand and saltbrush, to be brought back next day a prisoner, disguised in a *peon's* lousy garments, willing to sign away half a continent to save his own precious neck. (He was thereafter hidden up the Brazos until November, when Houston sent him to President Andrew Jackson, who sent him home—his gift to the host country the introduction of *chicle,* or chewing gum.)

Santa Anna had divided his main army into three divisions: one to sweep the prairies below the Hill Line, one to sweep the coast, and his own to wipe out the settlements between. Houston was aware that if Santa Anna were known to be dead the several Mexican generals in Texas would act on their own, combine forces and wipe out his exhausted handful of volunteers who were farmers and not trained *soldados.* His post-San Jacinto strategy (after frightening his captive enough to cause him to order evacuation of the ceded territory) was to confuse these generals by keeping their commander-in-chief up the Brazos at Orizimbo Plantation (now Varner, the Hogg Plantation) until they crossed the Rio Grande with watchdog contingents of Texicans nipping at their heels.

13. Hackamore: Texanism for the Spanish *jaquima* (ha-kēy-mah: "j" is "h" in Spanish.) Mrs. Berlet, as well as Brooks, notes the sequestering of the cotton yarn.

14. One wood clock is the single item of personal property listed for John Atkinson in the 1840 *Census of the Republic of Texas,* made up from the tax lists. The clock now ticks in the home of Mrs. Mary Janice Brooks, El Campo, Texas.

one-man military escort.[15] En route back down the Colorado River, they passed a surveyor's crew laying out the ground for a new seat of government.

It was a matter of debate throughout the Republic as to whether a capitol should be built immediately below the Balcones—a challenge to the Comanchería—or kept in safety among the settlements, either at Houston's namesake town on Buffalo Bayou (where, as Texas' first president, he had paid off the Tonkawa for their help in the hour of need, each with ten pounds of war paint) or at Washington-on-the-Brazos, where the declaration of independence from Mexico had been signed on March 2nd, 1836, four days before the fall of the Alamo.

The site on the Colorado—to be called "Austin"—was the choice of the second President, Mirabeau Buonaparte Lamar. In his hatred of Indians, whom Houston loved, Lamar had as his slogan, "Let the sword do its work!"

Frederic and Harriet, interested as citizens, had, at the moment, no personal concern with whether the capital should be on the Bayou or the Colorado. Their immediate concern was with a field on Mill Creek where corn would be planted for their next season's bread, for, as Harriet said, "Every tub must stand on its own bottom."

---

15. From Minnie Ray Swofford's ms., "Our Bachman Ancestors and Our Kin." See fn. 2, Chapter 6, "Good Man Gone Wrong."

The Land Juggler

Beyond this be dragons!
—Notation at the edge of the known world on an ancient map.

Nuclei of the Republic of Texas were nineteen counties, each, but two, converted from a Mexican municipality, or "city state": each *ayuntamiento,* or seat of city government, becoming a county seat. Many held jurisdiction (on paper) over an immense territory. From Bastrop, mapped far up into the Comanche country, twenty-eight other counties would be created. Among the earliest was Travis, when it was decided in 1840 that a point on the Colorado River, previously called "Waterloo," should be the capital.

It was on March 13, 1841, while the government was being removed from the mosquito-ridden bog-spot on Buffalo Bayou called "Houston" to a moated and stockaded one-story structure high up the Colorado, that Frederic and Harriet welcomed their second son and first Texas-born, William A. Their political concern was the concern of all settlers—that Lamar's hostile policy toward Mexicans and Indians would undo Houston's presidential effort to bring about peace on every side. Riling the Plains Indians (which included the Comanches' Apache allies, the Kiowa) not only had stepped up frontier raids on settlers, it aided and abetted Mexico's resolve to retake the lost municipalities. Both John Pinkney Atkinson, eighteen, and his uncle, John Beneke Bachman, but three years older than he, sought to participate in this tit-for-tat action back and forth over the Rio Grande that was to go on through years and for both the effort would end in disappointment.

In June of 1841, as John Bachman, with the Santa Fe Expedition Lamar was sending to project Texas' claim over New Mexico, waited on Brushy Creek some twenty miles north of the new capital site, he came down of a fever and, as in 1839, when his hand was injured, had to forego action.[1] A presidential term was, by the Texas Constitution, two years, with an incumbent not to succeed himself, but it allowed Houston to succeed Lamar and he filled the office through its third period. In October of 1842, following General Adrian Woll's foray into Texas when San Antonio was seized then abandoned by the general's flashback into Mexico, John Pinkney came home one day to find his other uncle, William Gordon, a year younger than he, had gone from Bell's Landing to join a punitive expedition under General Alexander Somerville. He would have struck off to overtake the avengers had not older heads prevailed: to start out over the prairie alone would be to court Comanche ambush.[2] Houston's recall of Somerville, when the Texas side of the border was clear, aborted the expedition except for a foolhardy three hundred (not including William Gordon) who willfully plunged on to be captured the day after Christmas at Mier on the Mexican side of the Rio Grande. Thence they were marched afoot over the long road to a Mexico City prison with an escape attempt on the way causing every tenth man to be shot by Santa Anna's order.

Shortly after this episode, known historically as The Mier Expedition, Old John Bachman and his sons, perhaps because they had had enough of Texas for the time being, went home to Mississippi.[3]

Houston, believing central Europeans capable of becoming friends with the Indians (which it seemed Anglos would never do) and thinking that a buffer zone might end the Comanche-Kiowa raiding, invited immigration to colonize the hills beyond the Balcones Escarpment. In half his surmise he was correct—a peace treaty would be the neo-colonials' immediate objective, though they did not arrive until the end of his term and that of his successor, Dr. Anson Jones.[4] The result of his and the doctor's joint statesmanship was the Articles of Annexation—a contract that set Texas' lone star as twenty-eighth in the constellation of the burgeoning United States.[5]

---

1. Swofford: ms. "Our Bachman Ancestors and Their Kin."

2. John Brooks, Bellville, ms.

3. P. F. Cochran, Houston: notes.

4. Roemer: *Texas.*

5. Texas was not altogether welcomed into the Union: just as proponents of annexation wanted Texas as an ally, lest France or Spain attempt a territorial claim on the grounds of early exploration, so opponents wanted no involvement with Texas, lest it lead to war with such claimants. Overshadowing the

Sarah Ann Norman Jourdan, whom her parents addressed as "Sallie," was born during Houston's second term. That year (1843) saw Frederic, then thirty-two, launched on the career he was to follow through life—he became a land juggler. In February, he bought 247 acres on New Years Creek in Washington County from John Burleson for $600 cash. This was part of a half-league Mexican grant to Abner Kuykendall (deceased) who had been in Stephen F. Austin's Old Three Hundred.[6]

On November 14, 1846, Frederic sold this acreage to Henry V. Roberts for $500 cash.[7] It was one of the few times he would lose money on land. Between the two deals, he named another daughter, Amanda Bradford "for a little girl I used to know."[8]

The next year, on September 10, Frederic paid Samuel Kuykendall, a son of Abner, $1,000 cash for 533 acres on West Yegua (yáy-wâh) Creek in the Evergreen Community near the Old San Antonio Road.[9] The family then removed from Mill Creek to the Yegua, where peaks of cone-shaped cedars loomed as if pollarded by a thousand skilled gardeners.[10] With them they took two Blacks and sixty head of cattle, as well as a fifth daughter, Louisa Jane.[11]

A few months after the Jourdans made this move, John Beneke Bachman returned to Bell's Settlement with a wife, *née* Elizabeth Ritchie, and two small children. He had written previously, asking his nephew, John Pinkney, how the hunting was holding out. John Pinkney, who had claimed a single man's headright in Austin County, assured him that turkeys, deer and bear were plentiful as ever on Mill Creek and the Bernard River, that the cattle range was unchanged, and that the price of a cow and calf was eight dollars.[12] William Gordon, married to Caroline Green, was also planning to return to

---

international aspect of the question was the internal problem of admitting another slave state. A treaty with an independent nation required a two-thirds vote of the United States Senate, but a contract required only a majority vote. By December 29 of 1845, such a contract—The Articles of Annexation—was agreed on by the two republics. President Anson Jones spent the next months preparing Texas for statehood. (For the Articles, see Appendix A.)

6. Deed Record, Washington County, Book H, p. 57.

7. Deed Record, Washington County, Book G, p. 346.

8. Reminiscence, Amanda to writer.

9. Deed Record, Washington County, Book H, p. 409.

10. Red cedars, *Juniperus virginiana.*

11. Tax report, Washington County, 1847.

12. Letter (Swofford files) John Pinkney Atkinson to John Beneke Bachman.

Texas. Both brothers had unpaid claims for service to the republic.[13]

On April 21st of 1850 (San Jacinto Day), Harriet inscribed in her Bible the name of a third son, "John Bachman Jourdan." Her elder daughters, Harriet Emily and Mary Catherine, were grown into young ladies of sixteen and thirteen. George W. and William A. were fifteen and ten; Sallie was eight, Amanda five, and Louisa Jane (called Jennie) was three. They were a healthy lot, but the same could not be said of their mother. Harriet's feeling poorly prompted doctors to tell Frederic that if they went on living in the malarial miasma of the low country, he could not expect to have her with him much longer.[14]

Although high and dry above the juncture of Buffalo Creek with Keechi (keēch-eȳe) Creek in Leon County, the Jourdan's headright lay untouched for they had no intention of going out onto the open frontier where Comanches murdered parents and carried children away. A year before baby John was born they had deeded this trapezoid of land to Franklin B. Blackburn for $640 cash.[15] From this point, Frederic's buyings and swappings become so numerous one gets lost among the transcripts. Two records of particular interest deal with other than land:

> On September 3, 1849, M.B. Alexander and William Hunt witnessed the signing of a rather informally worded paper wherein M. and M.R. Banur gave over to Frederic all right to and interest in "a certain nigra (*sic.*) woman named Lucy," about twenty-five years old, then "in the hands of the Sheff (*sic.*) of Bastrop County."[16] No compensation is mentioned. On September 8, Frederic gave to Samuel Croft, Sr. a bill of sale for Lucy, cash payment $600, the transaction duly recorded next morning at 9 o'clock by the Bastrop County clerk.[17] Oddly, the paper of September 3 was not recorded until November 6, and

---

13. 2nd Class Certificate "B" No. 3980, May 23rd, 1853 (Texas State Archives) acknowledges the "Public Debt of the Late Republic of Texas . . . a Claim for Services in the Vasquez Campaign—amounting to Thirty-One 50/100." The claim is supported by statements from John Hutchens and John Pinkney Atkinson, sworn to in the court of John Beneke Bachman, Justice of the Peace, Austin County. Furthermore, William Gordon appoints T. H. Kuykendall as his agent to appear before the state auditor and collect the claim. (That so many men of status in a community should be involved for so small a sum speaks for the value of a dollar in Texas in the period following annexation.)

14. Talk in the family, repeated to writer by Jody, her mother.

15. Deed Record, Leon County (Centerville), Book A, p. 441.

16. Deed Record, Washington County (Brenham), Book I-J, p. 102.

17. Deed Record, Bastrop County, Book P, p. 370.

even then over in Washington County where Alexander appeared before the Chief Justice to declare that he did indeed see the Banurs sign "the fore-going bill of sale" (it reads instead like a quit claim) to which his and Hunt's signatures were affixed.

It is obvious that there was a stir-up about Lucy. One would like to know the story. Was she a run-away, captured by the Banurs and held till ferreted out by the sheriff? Had she been leased to them and they refused to surrender her when Frederic arranged a sale to Croft? Was a debt involved and she seized for payment? However impossible to answer, it is sure that a legal artifact affords as scant knowledge of the past as any other relic from a bygone era.

On March 14, 1851, Frederic and Harriet bought from William Brown and his wife, Elizabeth, 84 acres "with improvements" on the east bank of the Colorado in Bastrop County, near the Travis County line. They paid $534 cash.[18] To this they added, on June 18, for $2,000 cash, an adjoining 471 acres bought from William J. McClelland and wife, the entirety of 555 acres from a grant to Solomon Duty (deceased) by the Republic of Mexico.[19]

This spread, just inside Bastrop's western boundary, was close to where, in Travis County years back, a middle-aged Vermonter, Dr. John Webber, built a fort, bought from a neighbor a dusky young wife, Sylvia (familiarly called "Puss"), and hired an English tutor to educate their children.[20] A community growing up around the fort had become a string town on the road west to Austin called "Webberville." Since this was much nearer than the recycled *ayuntamiento,* Bastrop Town, the Jourdans would become a part of this general community.

Exactly eleven months after Frederic and Harriet's first purchase in Bastrop County, Gibson Kuykendall bought their Yegua Creek holding for $1,106 cash.[21] That to remove from this place to the

---

18. Deed Record, Bastrop County, Book H, p. 42.

19. Deed Record, Bastrop County, Book H, p. 47.

20. Noah Smithwick, in *Evolution of a State,* gives the Webber story. Dr. Andrew Forest Muir, who worked with the writer in gathering data on Texas Blacks, left in his notes (Rice University Library, Research Department) many names of Webber descendants along the Rio Grande, also the finding that Mrs. Webber (Sylvia, or "Puss") drew a pension for her husband's service in the United States Army.

21. Deed Record, Washington County, Book K, p. 121. "Black Birders" were illicit slave runners not averse to seizing free persons of color and trading them where they were unknown. (James W. Fannin, in command of the Texas army at Goliad, was a Black Birder.) That Dr. Webber was justified in fearing

Colorado River bank might engender a greater hazard than to remain where they were was not thought of, nevertheless, the household was awakened at midnight in March, 1852, by a rider warning of a flood on its way down. The house already was surrounded by water which continued to rise as the family, Blacks included, waded to a log corn crib on higher ground. They were cramped for two days in this one-room structure with a dirt floor, waiting for the river to withdraw.[22]

Obviously the new buy was no place for Harriet. Not many days after the flood the older girls were on horseback, driving a handful of cattle (more would be rounded up later) up-river and west to a house Frederic had contracted for on Walnut Creek, a tributary of the Colorado.[23] The house stood where a road crossed the creek and ran seven miles south to Austin where a new capitol building, a three-story stone structure with a pillared portico and a modest dome, was two years old. On the other side of the creek, within shouting distance, was (and had been since immemorial) a Tonkawa encampment.[24] Even though the area was in easy reach of the escarpment sculpturing the Hill Line, it seemed safe enough due to the Annexation Contract: the United States would handle the Comanche problem.

---

for his children is evidenced by the tax record of 1850 where they are listed as slaves he failed to report—a record later corrected. Although Smithwick says he removed to Mexico, old maps show the states of Coahuila, Nueva Leon and Tamaulipas extending to the Nueces and Dr. Muir found the vastly extended Webber family presently in that part of what is now Texas.

22. Search of records by Mrs. Frances Moore, librarian, Austin and Travis County Collection, Austin Public Library, December 16, 1976, shows no great flooding of the Colorado River between 1852 and 1869. (See Appendix B for Grandfather Giles' description of the 1852 flood and other statements concerning the family's move to Walnut.)

23. The 1852 tax roll for Bastrop lists nothing but land, indicating that range cattle may have been too scattered or lost in the flood to allow for counting.

24. Amanda, reminiscing to Eugene, "Our nearest neighbors were Tonkawa Indians."

Below the Balcones

A patient Sculptor chiseled this when time was young and earth, a neophyte, drowned in baptismal waters.
—Lilian White Spencer, *Arrowheads*

Life was eerie on Walnut, though not without its charm. Wolves howled at night, echoed by owls, whippoorwills and momentarily waking mockingbirds. Through the day prairie chickens and turkey gobblers drummed and strutted, mimicked by quail. Doves startled from everywhere, their wings whistling, as deer skimmed over the grasses, bounding for the joy of it. Watching them all, turkey buzzards circled the blue above, waiting for leavings when the wolves had done with their prey.

When there was a cloudburst, the run-off roared over the household spring at the head of a little box canyon, its fall all but drowning thunder of the storm as it roared on down to the creek crossing. Chapparal cocks, almost tame (perhaps from association with the Indians, who revered them as "the Medicine Bird") ran ahead on the trail or flitted to a twig and waited, nearly in reach, to watch one's passing.[1]

It was at the spring, the Jourdans were told, that fifteen-year-old James Boyce was found in 1841, his body mutilated, the mule he had ridden gone. That was when the Lipan-Apaches were attacking the

---

1. Chapparal cock: "Among charms and amulets, objects presumed to prevent evil by their own innate power," Col. Richard Irving Dodge in *Our Wild Indians,* speaks particularly of feathers from the *paisano,* or roadrunner, known to the Plains tribes as "the Medicine Bird." No man would start on any enterprise without at least one of these precious quills on his person, while whole skins were hung over lodge doors to keep out henchmen of the Bad God. (Atkinson, *Indians of the Southwest,* p. 244.)

Chapparal: tight brush growth—often thorny; *paisano*: countryman.

Tonkawa and the Tonks fled to Austin where young James had been headed. He was the son of Elizabeth and Aaron Boyce, over on Gilliland; each morning the family refilled holes wolves dug each night trying to get at his remains in a plot marked off near the house for a cemetery.[2]

Hearing this and other stories at their dining table in the northwest end of a dog-trot (hall), the newcomers sometimes shivered in horror over meals brought in from the outdoor kitchen—a log cabin where coals continually glowed on a hearth above which game, pork or beef hung roasting.[3] Many foods were in abundance: vegetables, fruits and eggs, but Texas' wild cattle were not up to southern cowpen standards, making milk and butter scarce. Scant baking could be done other than yams (sweet potatoes) buried in the ashes or candied in their own syrup in a Dutch oven, with the mainstay, cornbread, always in a companion "oven."

Leon Hale, the *Houston Post,* May 13, 1979, quotes from a letter written in 1848 by Mrs. Elise Willrich, La Grange, to her father in Germany, anent the open hearth baking problem: "It is troublesome to make bread in this way, for it has to be done three times daily, and invariably at the moment when the hearth is already filled with pots and pans, and to these must be added two pots of bread with their lids filled with red-hot firebrands which give off a terrific heat."[4]

The new house on Walnut was modest enough, built, like the original capitol, of heavy lumber hauled up by six-and-eight-yoke ox-teams from the sawmill at Bastrop's pineries. There were two front rooms with fireplaces, and, after Frederic put them on, shedrooms at the back, on either side of the dog-trot hall. He would add a gallery across the front where the family could spread pallets and sleep in the summertime.[5]

Formal entertainment was perforce homemade. They might

---

2. The Boyce family cemetery—where James' grave is marked—is now within the property of Carl and Mary Jo Anderson, parents of Jenifer Bird, who, with her husband, John Bird, now lives on Gregg Lane. For longhorns as milk cows see fn. 6, "Interlude Major." Gilliland Creek, named for its first settler, James Gilliland, a Methodist preacher killed on Brushy Creek in 1839 when a Ranger contingent under General Edward Burleson was driven back by a large band of Comanches.

3. Outdoor kitchen: a customary arrangement of plantation houses.

4. Dutch oven: a thick and heavy, rather deep iron pan with a handle and rimmed lid; the rimmed lid was also a pan in which hot coals were heaped; thus, with fire below and above, the pan became a bake oven. Mrs. Willrich's letter is in the La Grange Museum.

5. A letter from John Lafayette Alford to Eugene, November 11, 1839,

invite settlers from miles around and get George W. (called "Wash") Cazine, Aunt Betsy Cazine's Black fiddler, to play for a square dance. But often as not they made do with a charade where some cut-up, donning a ratty old rawhide hat that spread far beyond his ears, would seat himself on a rawhide-bottom chair in the kitchen, drag a broom handle between his feet, and, vibrating it against the puncheon floor, sing to the rhythm thus produced (with appropriate gestures), the Brandy song:

> Chorus:
> Texans are Texans, anyway you fix 'em;
> Brandy's brandy, anyway you mix 'em;
> An' good good people are all in bed
> When the Devil gets a workin' in a Texan's head!

Starting in a deep bass (bullfrog reproduction), the singer adapted the words to his geographical presence, as:

> Come all you Louisiana gals an' listen to my noise
> But don't you marry any o' the Texas boys!

The tune would shift to a whippoorwill variation:

> They'll take you down to San Jac Hill
> An' that you know's agin yo' will!

The last line, repeated, was drawled:

> An' that you know's agin yo' will
> An' *that* you know's . . . *agin* . . . *yo'* . . . *will*!!

Such a performance was known as "The Broom Scratch."[6]

It was rough humor and crude music but all present joined in,

---

says there was no gallery originally. The habit of sleeping outside in warm weather was well established in Texas.

6. Broom Scratch: this version reproduced from memory by Mrs. Julia Jansen Hild, who had it in her childhood on Galveston Island from Mrs. Sallie Marie Trehan. Mrs. Trehan first heard the skit as a girl, when she came to New Orleans from France. The writer's daughter, Rosemary, researching files of Texas folklore in high school, came on versions naming several other states in place of Louisiana.

clapping and foot-stamping. It suited the times as well as the melodious harmonica that could be whisked from a pocket at a hint for an offering like "The Yellow Rose of Texas" or "My Darling Clementine."

The Tonkawa would come up from the creek to watch these wondrous doings and to accord, in turn, marvels of their own wandering world. They came sometimes to say swimming holes and nut groves must be cleared quickly for crickets had ceased their fiddling—that silence in the creek bottom meant water was coming down. This was no idle warning. On the brightest sunny day he who failed to heed, turning in a grapevine swing, might face a wall of water rolling soundlessly down to plunge him against Walnut's rock ledges—thence to be swept, swirling, to the Colorado and on and on, past Webberville and Bastrop and farther communities, to the sea.[7]

August of 1852 found Frederic in the Bastrop courthouse defending his title to the 555 acres on the east bank of the Colorado "with its meanders" against one J. Towns, intervenor on behalf of "the unknown heirs of M. Duty, deceased" (the brother of Solomon Duty, deceased). A jury of "twelve good and lawful men" (to quote the court record) ruled in Frederic's favor, ending anxiety on that score.[8]

Floods or no floods, on May 18, 1855, Frederic increased his holding by 50 acres from the adjacent Martin Wells league for which he paid H. B. Wells $500—"good and lawful money" (the deed says) "of the United States."[9] His fourth son, Zackie (full name Zachariah Pannell Jourdan), later to be known as "Z. P.," was then a year old, and his eldest, nineteen-year-old George W., wed to Gibson Kuykendall's sixteen-year-old daughter, Zulima C. (Catherine?), the license duly recorded on July 16, 1854, in the Burleson County courthouse at Caldwell.[10]

There is nothing to say where or how this young couple spent the brief years of their marriage, or how Zulima and her two young children died. She was the granddaughter of the widow, Elizabeth Kuykendall, who came into Bell's Settlement in 1829 with two sons

---

7. The Balcones Escarpment rises abruptly 450 feet in Central Texas; down the Colorado gorge above Austin water from the rains back in the hills rushes with a passion and, without warning, fills creeks below the Hill Line.

8. Bastrop County Record, Deed Book H, p. 426.

9. Bastrop County Record, Deed Book P, p. 532.

10. Zulima: The Memoirs mention only that George lost "a wife and two children"—no further data than the marriage record has been found.

and three daughters: Catherine, Martha, and Zulima.[11] The older Zulima married James Hodge, but her sisters Catherine and Martha married Abner's sons, Barzilia and Gibson. According to the deed of sale Frederic gave him at that time, Gibson was living in Burleson County in 1852.[12] The 1850 census gives a Zulima C. as Martha's first child, and his post office as Evergreen. On November 26, 1856, George's sister, Harriett Emily, married Baldy Samuel Alford, a Bastrop man. A year and three months later, Mary Catherine married Baldy's older brother, Will, and in May of 1858 George was married a second time, to Lucy Payton, in Bastrop.

The early years on Walnut were innocent, the halcyon period in the lives of Frederic and Harriet's children—the time when young love was burgeoning. Tragedy, except for George's trouble that seemed past when he remarried, cast as little shadow on their thoughts as did the circling turkey buzzards. The notion that larger vultures were circling unseen—vultures of a war to come suddenly as a wall of water down Walnut or an April thunderstorm—haunted only their parents. Even for them, the world beyond the Mississippi seemed far away from where the Tonkawa gathered each season when black walnuts and pecans were ready for thrashing. When the Tonks were not there, one's senses got lost amid the scenic distractions and the great orchestration of nature, though less so down among the chirruping crickets in the bottom land than out on the vast loneliness of the prairie.

Six weeks after George remarried, his mother bore her tenth child, the baby Julie, inscribed in the Bible as "Julia Elizabeth." On December 16, 1858, Harriet and Frederic sold 506 acres of their Duty-Wells tract to Joseph L. Hamer for $7,500, $2,000 down, the remainder due without interest January 1st, 1861.[12] In the afternoon of February 19, 1859, they closed a deal with Henry N. and Sarah Burditt: $4,000 cash for 492½ acres of creek and pasture land inclusive of the house they were living in, thus making the place their homestead.[13]

---

11. Betty McCosky's ms. in the State Archives lists the children of Mrs. Elizabeth K. Kuykendall, who received a league grant in Austin's Colony on Mill Creek in 1828, as two sons and three daughters. The 1850 Census lists Gibson Kuykendall and wife, Martha, as living in Burleson County, but receiving their mail at Evergreen. Obviously, they were neighbors of Frederic and Harriet. Their eldest child, aged twelve, would have been sixteen in 1854 when George Jourdan was nineteen. Although this child is given as "Julima, male," and the family as 'Kinkindall," it has to be the lost Zulima.

12. Bastrop County Record, Deed Book L, p. 164.

13. Homestead Law enacted by Congress, the Republic of Texas, 1839:

With but 99 acres left in his name on the Colorado, Frederic would register his cattle brand at the courthouse in Austin and bring four hundred cows up to Walnut.[14] He need no longer ride many miles to teach school or hold office as justice of the peace down Webberville way.[15] His mind was by this time on cotton as well as cows for he knew that, underneath the prairie grasses, the untouched black earth lay rich with promise.

Frederic's Brand

(ear crop and notch)
Book of Brands, Travis County record

> The Homestead of a family shall be . . . protected from forced sale, for the payment of all debts except for the purchase price . . . or for work and material used in constructing improvements thereon . . . any temporary renting of the homestead shall not change the character of the same when no other homestead has been acquired.

Since many persons coming to settle in Travis County—among them the Wallace Maxwells (Appendix C) leased land until they found a satisfactory purchase, it is here assumed that Frederic leased the Burditt place until he was able to dispose of all his Bastrop property except 99 acres.

14. This number of cows given on the Bastrop County tax roll, 1958.

15. That Frederic taught school at Webberville was related to the writer by Jody, who had it from family discussions when she lived with her parents-in-law in the house on Walnut. He is mentioned as justice of the peace, "Precinct 9," in the *Bastrop Advertiser,* Saturday, August 14, 1858: Richardson Files, Lyndon B. Johnson Library, Austin.

THE BROOM-SCRATCH

This version reproduced from memory by Mrs. Julia Jansen Hild who had it in childhood at Galveston from Mrs. Sallie Marie Trehan, then in her late seventies, who first heard the skit when, as a girl, she went to New Orleans from France.

C-O-O-O-O-o-o-me! (bullfrog theme)
All you Loo-zy-ana gals
An' Listen to my no-ise!
But don't you go to marry
The Tex-an boy-s! (ominous head shakings)

They'll take you down (tone of deep dark warning)
To San Jac Hill*
An' that, you know's
Agin yo' will. (whip'o will theme)

An' that you know's (whip'o will theme)
Agin yo' will!

Chorus: (very fast)
Tex-an're Tex-ans (stirring motion)
Anyway you fix 'em;
Brandy's brandy
Anyway you mix 'em:

Good, good people (teeth gritted, eyes screwed down,
Are all in bed hands clenched on broomstick and
When the Devil gits a-workin' roll of stick highly accelerated)
In the Tex-ans' head.**

When they go to milk a cow
They milk it in a go'ard;***
They set it in a corner
An' kiver it with a bo'ard:
Some gits little
An' some gits none—
An' that's the way
Them Tex-ans done!

An' that's the way
Them Tex-ans done!

Chorus:

When they go a-co'tin— (courting)
The way they dress! (ogle eyes)
Ole blue huntin' coat
Without any ves'! (vest)
An' ole dyed pants (expression of horrified disapproval)
They wear the year round—
But the Tex-an's hat's
Broad-brimmed an' crowned!

The Tex-an's hat's
Broad-brimmed an' crowned! (grand flourish)

Chorus:

*San Jac—San Jacinto—where the slight rise, insignificant to Texans would be definitely noticeable to tidewater Louisiana.
**Possible reference to Santa Ana's fatal siesta on the 21st of April, 1836.
***Early Texas kitchens had: "a gourd of sour milk or curds—clabber cheese which, in the absence of flour or meal was sliced for clabber " 'bread' "; a gourd of sweet milk and a gourd of eggs.

Bullfrog Theme

The Distant Tocsin*

If the cotton states shall be satisfied that they can do better out of the Union than in it, we insist on letting them go in peace . . . Whenever a considerable section of our Union shall deliberately resolve to go out, we shall resist all coercive measures designed to keep them in. We hope never to live in a republic where one section is pinned to the residue by bayonets.

—Horace Greely in *The New York Times*

Old John Bachman visited Texas again in his eighty-fifth year. He died at Harriet's home on the 23rd of January, 1860.[1] What his grandchildren would remember best about him was that when he became excited he lapsed into what was to them an incomprehensible jargon, the more bewildering because their mother's speech had no tinge of accent.[2] On February 2nd, Frederic contracted with Minus Burditt for 815 acres of prairie land adjoining that he had bought from Sarah and Newell Burditt, together with 65 acres of cedar or "mountain land," payment due a year from date of sale, was $8,885.[3]

In Austin, Sam Houston, who had shirt-sleeved the "Bantam Republic" from the Republic of the South (Mexico) and ramrodded it into the unwilling arms of the Republic of the North (the Union), was fitting a last string to his political bow.[4] It had been a decade since he

1. Letter from John Beneke Bachman, February 4, 1860, announced their father's death at Harriet's home to his sister, Elizabeth Bachman Riley—letter in the file of Minnie Ray Bachman Swofford, Lubbock, Texas. The letter states that the old gentleman "is buried near Austin," leading one to assume that his grave is in the Fiskville Pioneer Cemetery on a road off Rundberg Lane.

2. Jody's repetition of talk heard after her marriage into the family.

3. Deed Record, Travis County, Book O, p. 59. This contract allows for doubling the purchase price if not paid on time.

4. Shirt-sleeve diplomacy: threat of force without soft-soaping palaver.

* Tocsin: a call to arms.

left the Brazos and Colorado rivers for the Potomac, there to expend senatorial eloquence in an attempt to circumvent a cotton-and-tobacco plantocracy bent on either nationalizing slavery—extending it from the Atlantic to the Pacific—or tearing the Union apart.

Members of this plantocracy were moving out of a politically ambitious South onto Texas' fertile bottom lands where they would bide their time until they should be sufficient in number to out-vote Old Texans and bring the ex-president home from the senatorial role he was playing.[5]

Their time came in 1857 when they replaced Houston with Judge John Hemphill, a South Carolinian and chief justice of the Texas Supreme Court. The judge, driving from Austin to Washington, D.C., took with him in his carriage his young daughters, Henrietta and Theodora, whom he dropped off at Wilberforce University in Zenia, Ohio, a school just started (1836) "for the education of the colored children of southern planters."[6] According to public record, the girls' deceased mother was twenty-year-old "Sabina, a dark-complexioned slave woman," whom their father bought for $500 at Old Washington-on-the-Brazos when he held court there in the days of the Republic. Unlike the New England Yankee, John Webber, who married Sylvia, the judge did not marry Sabina—which was in keeping with the mores of his constituency.

In 1859, the year Frederic brought the main body of his cattle up from the Bastrop pasture, Houston was elected Texas' sixth governor in a hotly contested campaign.[7] People at large were at last aware of a danger Frederic and Harriet had foreseen and sought to escape—the danger of civil war.

Governor Houston flatly refused to call a secession convention at the behest of transplanted planters and Bourboncrats but on January 28, 1861, the convention was held anyway, its delegation far from representative of Texas as a whole.[8] The vote was 166 to 8 in favor of

---

5. In the controversy over the Kansas-Nebraska Act, designed to extend slavery into the West, Houston was "the only man from the South" to vote "no."

6. Notes from deposition by the president of Wilberforce University in a suit brought by Theodora Hemphill for control of her father's estate following the Civil War. Henrietta died at the school where the girls had a home for the duration. See fn. 5, Chapter Eleven, "Rose Cottage."

7. Frederic was on Bastrop County tax rolls through 1858; he appears on the Travis County tax roll in 1859, the year his brand was registered, and the year he paid for the new home.

8. Bourboncrat: a term Texas resurrects any time the South assumes

withdrawal from the Union.[9] With so much uncertainty in the air, Frederic's contract with Burditt was extended for a year.[10] Houston found himself the de facto head of a de facto nation: legally or illegally, Texas had resumed its independent status—a condition culminating on March 14, when, at demand of the secession convention, he refused to take an oath of allegiance to a "Davis government" hastily flung up at Montgomery, Alabama. He also refused an offer of military aid from Abraham Lincoln, the newly inaugurated President of the United States, and used the interim to clear Texas of all federal troops occupying Hill Line and West Texas forts, thus preventing conflict between them and a regime about to be formed under the secession movement. This accomplished as insurance against bloodshed within the borders of a country ring-bound on two sides by Napoleon's cast-off Louisiana and the open sea, on the other sides by Mexico below and the Comanche-Kiowa above, he let the more malleable lieutenant governor (whom he charged with having "slid

---

superiority, socially or politically. Its sources are varied but entangled. It is a historical fact that the Emperor Napoleon III (Louis), in one of France's spasmodic attempts to return to a Crown, favored the Confederacy over the Union as a prop to his scheme to mount Ferdinand Maximilian Joseph, brother of Franz Joseph, Archduke of Austria, as Emperor of Mexico, to which purpose his forces occupied the Mexican coast in 1861. It is also historical fact that the Napoleonic dynasty gained its titular foothold through marriage with royal blood. It may be only a folklore fantasy that Charleston, with a taste for the same sort of elevation, afforded life-long secret sanctuary to the last claimant to the French throne through Bourbon blood.

9. When the secession convention met at Austin in 1861, except for T. P. Hughes from Williamson County, all delegates were from north and northeast Texas, leaving San Antonio, Austin, and Fayette Prairie and eight "German counties" in the hills unrepresented. The German people Houston had invited to settle between the Anglos and the Comanche-Kiowa (the "Forty-Eighters") had left their homeland for Texas as a part of the European anti-monarchial movement. They had neither slaves nor plantations and no interest in the secession movement. Their loyalty was to the principle of republicanism.

The vote ran one hundred and sixty-six to eight in favor of leaving the Union and Texas was again, for a few days, an independent country. Those who cast the negative vote (favored by some historians as "the Eight Immortals") were, besides Hughes: W. H. Johnson of Lamar County, J. F. Johnson of Titus County, J. D. Rains of Wood County, A. P. Shuford of Wood County, J. W. Throckmorton of Collin County, L. H. Williams of Lamar County and G. W. Wright of Lamar County.

10. See fn. 16.

into office on my coattails'') slide on into the administrative chair while he delivered a farewell address in the words of savants regarding war since time immemorial: ''Texas,'' he mourned, ''must soon become a nation of old men and boys.''

General Robert E. Lee, in command of the sweep of forts through Texas to the Rio Grande, was on furlough at the time. He would not return. The emptied forts were an open invitation to the rejected Comanche-Kiowa for renewal of what would be known as their ''Forty Year War'' on Texas. Before Annexation, they had asked to become an integral part of the commonwealth and were rejected, not so much by Old Texans, it was said, as by newcomers who, like Lamar, had no respect for Indians.

On April 12th, South Carolinian Hot-Heads, not waiting for diplomatic action between a jerry-built Confederacy and the materially solid Union, believing all they had to do to win a war was to start one, besieged federal troops at Fort Sumter and took control of Charleston harbor. The holocaust was on.

The summons from beyond the Mississippi did not overplay its championship of slavery. Aware of an awakening Christian conscience on that score, the call to arms was, instead, for defense of Jeffersonian demoncracy and states' rights. Earliest Texicans to answer were the young unmarried men, among them the first Texas-born Jourdan, his sister's darling Billie. He went in November of '61 with the Travis Rifles into the Army of Tennessee.[11]

Late in the spring of '62, when Davis demanded that Houston's successor provide fifteen regiments from seven adjacent counties, the latter announced that he would impose a draft were there not enough volunteers to fill the quota.[12] By then death notices had considerably damped the ardor of idealistic ladies who sent the earliest contingents off with ''bonnie blue flags'' pieced from their silk dresses—except, of course, for the Red-Hots, who were turning to the wall portraits of men choosing to defend the principle of ''my country, one and undivided,'' and who were naming their girl babies ''Secessia.''

It was on May 12th, 1862, that George Jourdan left his wife Lucy, their newborn George W., Jr., their three-year-old James W., their two-year-old Frances Priscilla, and his place in the Capitol Guards to

---

11. William A. Jourdan, private, Company G, (Travis Rifles) 6th Regiment, Texas Volunteer Infantry, Army of Tennessee. CSA Records, Texas State Archives.

12. The seven counties: Anderson, Bastrop, Burnet, Burleson, Travis, Milam, and Washington.

join his younger brother's regiment.[13] Harriet Emily's husband, Baldy Alford, left the same day with a Bastrop Guard to join Terry's Rangers.[14]

Close of the second year of conflict found every Texas man under forty-five either in uniform or in some vital civic position, while government agents came to measure Frederic's and his neighbors' corncribs and determine how much provender each might keep and how much would be commandeered for the Confederate Army.[15] Where there were no Blacks to work the fields hard-won from prairie grasses the ground lay fallow. Where cattle knocked down fences and got out with mustangs and longhorns on the open range the fence rails were not relaid. But on November 20th, 1862, Frederic, with a "black family" of nine to depend on, paid to Minus Burditt the $8,885 he owed for 871 acres of land, thus expanding his holding to more than 1700 acres—a size to qualify him, when the war should be over and the sod broken, as a planter rather than a cattleman.[16]

Like everyone else's, his and Harriet's house was beginning to overflow with women and children. Harriet Emily, waiting with her five-year-old daughter (also named Harriet Emily, but called "Dude"), her two-year-old son, John Lafayette (called "Jack") and the baby, Baldy Amanda Belle, born after Baldy's departure, would be months in suspense before learning that he had died of pneumonia before he had been a year from home. The hymns of love Emily composed for her diary would be his memorial and her testimony to an abiding loneliness and a crumpling grief that she might not even know his grave.[17]

With every day life grew more austere. Gourds cut into cups and bowls replaced broken dishes, rawhide sandals the worn-out shoes.

---

13. George W. Jourdan, private, Company D, (Travis Rifles) 6th Regiment, Texas Volunteer Infantry, Army of Tennessee.

14. Baldy S. Alford, private, Company A (Terry's Rangers) 18th Brigade, Texas Volunteer Infantry.

15. Jody's repetition of reminiscences heard after her marriage in '73.

16. Deed Record, Travis County, Book P, p. 550—wherein Frederic is duly "enfoeffed" with the listed properties—a legalism harking back to medieval feudalism. (It is not to be supposed that he actually went out to the four corners of his new domain, flinging the earth in all directions and shouting that he was now taking possession!) "Black family": in common usage for a plantation's Black population.

17. Harriet Emily's Diary: a notebook in which she kept records and wrote her poems—left to Emily Swartz Rutland, a Kingsville, Texas, artist, by her mother, Baldy Amanda Belle Alford Swartz.

When the "Feds" blockaded the little Gulf ports white flour vanished, as did coffee and sugar. Make-believe coffee from parched okra seeds was stirred with "long sweetening"—either sorghum syrup or melted *piloncillos* (peél-on-seél-yos), the latter brought up from Mexico when planters began sending long "bull trains" down the ancient trail worn by missionary priests and *carteros* (Mexican freighters) to Brownsville and Matamoras.[18]

At Matamoras supplies (particularly medicines) from England and France (both covertly allied with the secession movement) could be picked up for the Confederacy. France, taking advantage of the overall situation, had a military force in Mexico City preparing to seat Maximilian Habsburg and his wife, Carlotta, on the throne of a toy empire.

The route the cargoes took was safe enough even though the Hill Country forts were unmanned and the Plains Indians had returned to their war with Texas. It lay well below the range of Comanche raids; besides, the Comanche-Kiowa were leary of Blacks. They would dub them "buffalo soldiers" because of their color and hair when the U. S. Army took Freedmen into post-war service and the Hill Line forts were reactivated.[19] The long bull trains snaked down through the Mustang Desert that was the Brush Country below San Antonio. Each load of cotton drawn by six or eight yoke of oxen was under control of experienced Black men who could break a pair to walk in unison by tying their tails together, cut a new wheel from brush timbers if a load broke down on the way—or shoot to kill if attacked.[20] Youngsters like

---

18. Jared Groce (see Note 5, Chapter One) according to Mrs. Berlet, in conversation with the writer, carried on a cotton commerce with Mexico by oxtrain, as a colonist. His Black *carteros* brought back so many silver pesos they were piled in cotton sacks on the gallery of Barnardo, his Brazos River dwelling. Aaron, the blacksmith, hammered some of this surplus silver into pitchers and platters for the family table. Though called "bull trains," the draft animals with their incurved horns, underslung bodies and knobby knees, were bovine eunuchs—steers.

19. See Note 8. A mystic element enters the Comanche's reaction to Blacks—the buffalo was the animal sacred to the sun, its death looked on as a sacrifice that the Indian might live; an elaborate ceremony preceded its slaughter.

20. Kenneth V. Porter, in the *Southwestern History Quarterly*, Vol. 53, No. 2, in relation to the antebellum Texas Black says:

> A slave holder would be unlikely to bring to such an exposed frontier a Negro who was not thoroughly trustworthy; the danger of his running away to the Mexicans or to the Indians themselves

Frederic's thirteen-year-old John were allowed to go along for the fun of it, wading out past the middle of the Rio Grande when they got there so as to boast they had "been on foreign soil."[21]

It was a far cry from the day when John, aged four, mounted behind his father, was discovered by his mother about to depart bareheaded in the sun for Webberville. The party of horsemen, held up while his sisters ransacked the house looking for his hat, were about to drop him in their impatience when he shouted, "Well, bring me a bonnet, then!" and one of the girls ran out to tie the strings of her slat sunbonnet under his chin—an episode the Mexican excursions would enable him to live down.[22]

It was at this time of the bull trains that a rumor got abroad, among countless other rumors, that some of the Hancock family in Austin were not speaking to others because somebody in the family who had gone farther on into Mexico had remained to dig for gold instead of coming home to join the army.[23]

On New Year's Day, 1863, Lincoln, who had neither a vested interest in slavery nor a familiarity with an enslaved people or their masters, issued his Proclamation of Emancipation. The proclamation would make little difference on Walnut and Gilliland where cotton would continue to be planted by hand and picked by hand and started on its long, jolting ride to the Rio Grande.

Not *piloncillos* nor the end of the whízzarówing wet northers, nor a carpet of wild flowers bursting into spring bloom could sweeten life for the Jourdans after word came that William A. was dead at Camp Butler, in Illinois, where George was still a prisoner of war.[24] In July,

---

was too great to be risked. Negroes on the Texas frontier were unlikely to be working in gangs under the overseer's lash; they were likely to be engaged in cattle herding, carpentry, corn harvesting, freighting—the same sort of thing their masters and other white men might be doing at the same time.

The *Diary of Reading V. Black* discusses the freighting industry and the importance of *piloncillos* in detail. Black's *Diary*: courtesy of Mrs. Margaret Ames Scott, Lafayette, Louisiana.

21. Reminiscence to the writer from her father.

22. Family lore repeated to the writer by Jody.

23. Lore in the Payton family.

24. Whíz-a-rów-ing (whiz-arrowing): a Texanism frequently used by the writer's father. A "norther" (north wind), from the Spanish *El Norte,* has been timed to drop the temperature a degree a second. A "wet norther" comes in with rain, often under a dark blue cloud, and is then a "blue norther." See Atkinson, *Texas Folklore,* Vol. 7, p. 75, 1928.

Houston died at his home in Huntsville, aged seventy, less mourned than he might have been were not tempers rising so high in a country divided from the beginning on the subject of secession and finally in shock over its consequences.

Early in August, Bellville sent word of Maria's death. She was sixty-three years old. She had had six sons and eight daughters. The three sons who lived to be grown: John Pinkney, Alexander and Solomon, were somewhere across the Mississippi. (Alexander would die two months later).[25] As more months passed, feelings intensified. Red-Hots reviled Austin and San Antonio as "damn Yankee towns" because each had a well-established contingent of antisecessionists, politically out-maneuvered but not to be annihilated like the company of men marching from the German colonies in the hills toward Mexico with intent to ship out for enlistment in the Union forces. They were overtaken by Red-Hots and massacred on the Nueces.[26] Old John Lohman who fled one European country for another to escape military service, only to find himself conscripted into an alien army, had sought Texas sanctuary in his middle age. Accused of sheltering a spy—a passing stranger—at Lohman's Crossing, upriver from Austin, and about to be hanged by an over-officious Red-Hot on a capital city street, he was snatched from the rope and sent home in safety with his bruised neck and his scars received in the battles of Leipsic and Waterloo.[27]

---

25. Texas State Archives: John Pinkney Atkinson, Company H, 26th Texas Cavalry, CSA; Solomon Atkinson, private, Company F, 16th Texas Infantry, CSA; Alexander Atkinson, Company F, Cavalry, Waul's Legion, CSA. John Pinkney, married to Martha Ann Cheek, died the day after he returned from the war, leaving two sons and a daughter.

26. A monument at Comfort, Kendall County, Texas, is said to be the only public memorial south of the imaginary Mason-Dixon Line honoring Union soldiers.

27. Lohman's Crossing, some thirty miles upriver from Austin, was submerged when a series of dams was built and the Lohman cemetery, like other family cemeteries, removed by federal order. The writer was present at this removal by invitation of John Lohman's granddaughter, Mrs. Herman West. John Henry Lohman, as the writer heard the story, reached Texas in middle age, married a sixteen-year-old girl and brought up a Texas family; the one surviving son, an elderly man, looking into the grave as his father's skeleton was exposed, exclaimed, "I want to see Pa's leg—they cut him on the leg with a sword (the "w" pronounced as Anglo-Saxon "swoerd") in the Battle of Leipsic!" The leg bones were passed up. One ankle showed a neat V-shaped notch, as if snipped out with scissors. Came a second demand, "I want to see Pa's head—they cut him on the head in the Battle of Waterloo!"

There were even those whose offense, in the eyes of the Red-Hots, was no more than never having owned a slave and not wanting one—unlike the man who stood on the capitol steps, with the secession convention in action, shouting "I ain't got a nigger! Gimmie a nigger, somebody, an' I'll fight for 'im same as anybody else, but I sure as hell ain't goin' to fight for nobody else's nigger!" "Yet this," comments Noah Smithwick of Webber's Prairie, who tells the story, "was the kind who had to do much of the fighting."[28]

Another winter brought back the northers—and word that George, returned to his regiment in a prisoner exchange—had been furloughed on sick leave and was coming home. Lucy, waiting with her little ones for news, would never hear when or how he died.[29] Tragedy seemed unceasing. Only Mary Catherine's husband, Will Alford, on detail within the state, was left of the family men who had gone to war.[30] All that kept life bearable was the sharing of misery.

At Corn Hill, near Merrilltown, some thirty miles north of Austin, where they had lived since their move from the Colorado after the great flood of 1862, Edward Stanhope Giles and his wife, Nancy Caroline, grieved for their son, Lewis Lucullus, who had gone at the war's beginning with Terry's Rangers and had died on its first Christmas day of wounds received at Mumfordsville, near Gatlinburg, in their old home state of Tennessee.[31] They had come to Texas in '49 with their eldest child, Marguerite, whose husband, Wade Henry, a blacksmith, saw Austin, capital of a new republic, as the place of opportunity.[32]

Up came the skull showing a deep dent across the crown.

Looking on, watching a left-handed grave-digger shovel the bones up from the dust to a right-handed assistant who, mounted in the hearse, was reassembling the bones in a thin pine box (perhaps a third the size of an ordinary casket), one could only hope the poor refugee from strife might know no awkwardness, come the resurrection and Judgement Day!

28. Noah Smithwick in *The Evolution of a State.*

29. The record in the Texas State Archives is confused; he seems never to have been heard of after starting home on sick leave.

30. Notes from CSA Records: Will Alford, private, 4th Battalion Cavalry, TST. Mustered in August 15, 1863, Travis County, for six months. Served one month, sixteen days at $12 per month plus $19.40 for use of horse, arms, etc. Company station at Liberty; scout of twelve men under Corporal Goforth from Camp Terry to Walnut Creek, Llano County, after Indians. TST reorganized and Alford mustered in March 16, 1864. Detailed to haul cotton.

31. Lewis L. Giles, private, Company D (Travis Rifles) 26th Brigade, Army of Tennessee.

32. Data from Giles Papers; Obituaries.

Unlike Frederic, who would have avoided the war if he could, and who was listed with the anti-secessionists, Ed Giles gloried in the conflict, at the same time praying for preservation of his other sons, William Leroy (called "Dock") and Valerius Cincinnatus (called "Val"). An avid reader both of the Scriptures and of classical history, Ed had named his boys for ancient statesmen and heroes (one buried in childhood had been Calvin Lycurgus).[33] The dogs left guarding the premises when Val rode off to fight for a dream called the Confederacy were Cassius, Brutus and Caesar.[34]

A father might take comfort in such words as those of the poet, Macauley: "How can man die better than in facing fearful odds for the ashes of his fathers and the temples of his gods?" but ashes of glory did little to assuage the hurt of Mollie Mc____, whose mother feared for her life or her sanity in the passion of her grief over Lew's death. (Lew's letters home make no mention of this girl. William Leroy, in a letter to his sister, Elizabeth Land, expresses surprise to hear of a romantic attachment.[35] Perhaps rather than an "understanding" with Lew, it was an overwhelming sense of loss and helplessness in the grip of an incomprehensible disaster that brought on Mollie's seizures of an emotional kind not uncommon in time of war, particularly among sensitive young women.)

Lew's father wrote of his sorrow to his remaining sons but expressed his pride in their patriotism. He wrote also of "that abolition hole, Austin," where ten dollars "of our Confederate money, our whole and soul dependence," were worth but one dollar in gold.[36]

Early in March of 1865, Frederic, smooth-shaven as always, a trifle grim under his high beaver hat, met one of the Hancock ladies in town. She paused to inquire his opinion of a likely Black boy (aged nineteen) for whom she had a fancy. She could get the boy for $1500, but in gold, not in Confederate script. Answering what he thought rather than what she asked, Frederic said she had best hold on to her gold, that the South was going to lose the war. She turned from him in disdain and closed the deal, as if by confirmation of faith she could

---

33. A sister, Eliza, was also buried in Tennessee.

34. Laswell, in Val Giles' Memoirs—*Rags and Hope*.

35. Letter, February 15, 1862, Camp H. E. McCullough.

36. Letter from Edward and Nancy Giles to their son, William Leroy, August 20, 1863.

LEWIS LUCULLUS GILES

NANCY CAROLINE Jackson Giles.

confute reality.[37]

In days, she had neither boy nor gold, for on April 9 General Robert E. Lee tendered his sword to General Ulysses S. Grant on the steps of Appomattox Courthouse less than a hundred miles from Richmond, capital of the Confederacy.

The long nightmare was over.

---

37. Family lore repeated by Jody.

WILLIAM A. JOURDAN

in Confederate uniform.

The Break-Up

Like the leaves of the forest when summer is green,
That host with their banners at sunset is seen;
Like the leaves of the forest when autumn hath blown,
That host on the morrow lay withered and strown.
—Lord Byron, "The Destruction of Sennacherib."

No mantle of glory fell on young Texicans who rode across the Mississippi in the fall of 1861, only mud and cold and hunger. Survivors getting back home as best they could in the spring of 1865 found a purple pall of mourning over the land and spiritual wounds that would scar generations to come. Nineteen-year-old Amanda, gentle and patient, married Daniel Rufus Payton a month to the day from Appomattox and he took her to live with his widowed mother, Phoebe Hancock Payton Forehand, in the Hog-Eye Community, Bastrop County.[1]

It would not be the life they had planned, nor was he the man she had first known and planned to marry. In his thirties, no idealistic boy, he had sent his army payments home for his future, only to find that his family, in their burning belief that the South must win, had invested his money in slaves that were become Freedmen.[2] Disappointment added to ill health (he had contracted tuberculosis) gave impetus to a naturally reckless spirit. His brother-in-law, John Jourdan, speaking in memoriam years later, would say, "Dan Payton

---

1. Appomattox: April 9, 1865. Hog-Eye: a dance form (hoe-down), the favorite of a Black fiddler who officiated at local play parties in a community now known as Elgin. Was it George Cazine? Phoebe Hancock Payton married Martin Forehand on August 25, 1853. He died within a year: note from Isabelle Dubar's files. Her first husband, Wesley Payton, died in 1848.

2. Family lore.

was as fine a young man as there was in the county before the Civil War ruined him."[3]

Between Appomattox and Amanda's wedding, Lincoln died. A telegraph wire newly invented, reached to East Texas. Word of the assassination, brought to Austin by runners, was derided as a scarehead until confirmed. Then, as bombardment of Fort Sumter had been the opening act in an insanity of civil war, so would the president's assassination be the opening act in a stupidity of reconstruction. A southern hand had held the gun.[4]

As vindictive measures got underway in Washington, D.C., Austin's secession movers and shakers were thrown into hysteria. On June 10th, the ailing governor, Pendleton Murrah (he was said to be dying of tuberculosis, and did die within weeks) fled Austin for San Antonio, en route to Maximilian, who would be crowned on June 11th. A horseback caravan accompanied the governor. Among the refugees, Judge Seaborn Graham Sneed and his wife, Mirinda, were abandoning a near-complete slave-built mansion at St. Elmo, five miles across the river from the capitol. The windows had been paned with glass ordered from Paris but the intended elegance of columns, front and back, would never leave the drawing board.[5]

On the 11th, looters at the state treasury were thwarted by Shelby's Raiders—a Confederate cavalry unit still in

3. Conversation heard by writer as a child.

4. Last gasp of southern madness was by John Wilkes Booth, twenty-six-year-old Maryland actor who thought to be acclaimed for his dramatic gesture.

As a teen-ager, the writer was taken to visit this house. Pieces of pale green glass, broken out by hail, still lay in the yard. The resident Sneed daughter explained that windows on this side had been replaced by ordinary glass, whereas that in the rest of the house was imported. Jody (the writer's mother), who was Mirinda Sneed's sister's child and the judge's ward after her parents' death, was living with the Sneeds when the post-Civil War exodus took place. The detailed contract for building the Sneed mansion appears in Deed Book L, p. 141, Travis County Records. Columns are not mentioned, but it was Jody's understanding that they were to be added, front and back, at a later time. It is interesting to note that one of the contractors was, or became, a Sneed son-in-law, and that, as told by Fannie M. Wilcox, Judge Seaborn Sneed's great-granddaughter (ms. Austin and Travis County Collection, Austin Public Library), the old gentleman had one stair tread made removable as a secret hold for his ready cash.

Cost of building the house was $13,000. It passed out of family hands with the death of the last resident granddaughter April 17, 1939, sale price $8,000, and is now (1978) up for sale, with 400 acres of land adjacent. The asking price is $10,000 per acre.

uniform—dashing through town on the way to proffer their services to Maximilian. It is said that they paused midway between Eagle Pass and Piedras Negras to bury their company flag and the plume from Shelby's hat in the waters of the Rio Grande.[6]

Texas was left in political chaos. The star that had shone with its twin on a blue field, symbolizing the conjoined states of Texas-and-Coahuila, that had been snatched at San Jacinto and emblazoned alone on the tricolor of an independent republic, that had twinkled for a decade and a half in the Union constellation and for less than half a decade in the southern Cross of St. Andrew, was in eclipse.

Far down in the tip of Texas, Lt. William Leroy Giles, quartermaster (Colonel Benavides' Regiment, Captain M. D. Wilson's Company) wrote his sister, Elizabeth Land, a letter on March 1st, 1865, headed "Camp Rancho de la Palmetto (*sic.*)":[7]

> Not a word from Val, poor fellow. Great God, I do not know how he has stood seven months in a Federal prison. Four months like to have killed me. Well the war still drags its weary length along. All this fuss among the Yankees was purposely manufactured by Mr. Seward for the Yankee draft market.

"Dock" Giles had left a job as Travis County's tax assessor and collec-

---

6. Benito Juarez (wâh-rez), an Indian patriot, was twice president of Mexico (1858-1863 and 1867-1872). The French incursion was between his terms—a venture abandoned, due to U. S. protest, after Maximilian's death. In the Centennial year of Juarez' death (1973) various countries were invited to send urns of earth, each from the grave of a national hero, to be interred in his grave. The United States sent earth from the graves of Washington and Lincoln. The writer was present in Saltillo, Coahuila, with a granddaughter in school during that summer.

7. According to botanical experts, the so-called "palms" of this area are not palms, but a tropical desert growth found elsewhere in the United States only in the tip of Florida. The mistake in identity accounts for *Rio de las Palmas* as Spain's name for the Rio Grande at its mouth. This river was also "discovered" and named *El Rio Bravo* in its middle course. It was as "the river of the palms" that La Salle sought it in his misguided tour west of the Mississippi; hence the Bourbon Lilies in the six flags of Texas.

tor to enter the army and he was sensitive to the economic aspect of the country.[8]

In a letter written previously from somewhere in the South (only a fragment remains) he had poured out his soul, saying that patriotism and love of country were given over to the hydra-headed monster "speculation," which had led to such inflated prices that an ordinary pair of boots cost $200, a common pair of jeans cost $200, a pair of pants $50 to $100, a pair of brogans $50 to $70.[9]

Whatever his personal opinion of the economy and the war, Dock would continue the monotony of rationing coffee, sugar, vinegar, candles, soap, lard (items obtainable from across the border) until May 13th, when a detachment under the colonel's brother, Captain Christobal Benavides, attacked four hundred Federals marching up from Port Isabel—killing thirty and winning the battle before the luckless intruders were able to make it understood that the war was over and had been over more than a month![10]

This postscript to the Civil War ended with the regiment's taking over the customs house and replacing with gold all the Confederate paper they were holding in payment of their service on the border: script from "the Davis government" good now for nothing more than souvenirs—or to kindle a kitchen fire.

On the 19th of June, General Gordon Granger, having landed at Galveston the day before with fourteen hundred men, read Lincoln's Emancipation Proclamation two years and six months late. Thus he started his tour of duty to inform Texas Blacks of what they already knew—that they were free of bondage. Later, they would celebrate the

---

8. A scrap of paper exists showing that W. L. Giles held this position.

9. Brogans: Scottish term for heavy shoes.

10. Christobal Benavides (the Colonel's grandson) gave this information to the *Eagle Pass Guide* April 19, 1956:

> The last battle of the Civil War . . . was fought near the old battle ground of Palo Alto. The charge against the Federals was led by (my grandfather) Captain Christobal M. Benavides of Eagle Pass. The Federals were driven back in a running fight in which thirty Union soldiers were killed and five Confederates wounded . . . Commander of the regiment was Col. Santos Benavides, Christobal's brother. Still another Benavides brother, Refugio, was captain of another company. The unit was known as The Benavides Regiment.

Several rationing orders in Dock's neat script are among the Giles Papers. The customs house incident is cited historically.

day as "Juneteenth." Just then what they termed "The Break-Up" posed problems too serious for jubilee. Riding from plantation to plantation, harbingers of freedom found that many whom they addressed preferred to remain where they were, at least for the time being.

On March 2nd, two years later (ironically, on Texas Independence Day) the federal government, subscribing to an angry North that would never comprehend behavior trends in a society accustomed, whether slave or free, to think in terms of gentry, common folk and trash, put Texas, along with Louisiana, under martial law and sent General Phil Sheridan (who had been active in crushing the South) over to enforce the northern concept of democracy—the idea that everyone is socially equal, regardless of individual differences.[11] It confused the general to find that Black people, who knew full well whose non-black blood mingled with African blood in their veins, saw no social percentage in the "bottom rail on top" idea and were not lured to the polls en masse while the Iron Clad Oath he was sent to impose debarred Confederate veterans from voting.[12]

The Texas star might be in eclipse—its spirit was by no means

---

11. March 2nd: The Texas Declaration of Independence from Mexico is handsomely framed in the capitol rotunda, companion to similarly framed Articles of Secession.

12. The origin of this phrase is perhaps an ancient proverb (quoted in the *Houston Post,* June 6th, 1976, as Zulu): "The wheel keeps turning—the bottom becomes the top, the top the bottom."

The Iron Clad Oath: *Harpers Encyclopedia of United States History,* Vol. 7, Harpers, New York, 1902:

> I do solemnly swear or affirm that I have never voluntarily borne arms against the United States since I have been a citizen thereof; that I have voluntarily given no aid, counsel or encouragement to persons engaged in armed hostility thereto; that I have never sought nor accepted, nor attempted to exercise the functions of any office whatever, under any authority or pretended authority hostile to the United States; that I have not yielded voluntary support to any pretended government hostile or inimical thereto. And I do further swear or affirm that to the best of my knowledge and ability, I will support and defend the Constitution of the United States against all enemies, foreign and domestic; that I will bear true faith and obedience to the same; that I take this obligation freely, without any mental reservation, and that I will faithfully discharge the duties of the office on which I am about to enter, so help me God.

quenched. When the federal military encamped at Fiskville and a contingent rode out to measure Frederic's corn cribs and estimate the content (just as Confederate conscriptors had done), Sallie, sprightliest of the middle three Jourdan girls (herself, Amanda and Jane) sat in a window singing "The Bonnie Blue Flag." Had Union soldiery been the monsters Jane thought (she would call them "blue-bellied Yankees" till she died) it might have fared ill with the young lady (there were those who marveled that the house was not burned over her head), but the blue-coated soldiery went quietly about their work and paid for what their horses ate.[13]

Frederic's reaction, though he had no love for conscription or martial law, was totally unlike that of Mrs. Mary Ann Rose, mistress of Buena Vista Plantation, farther eastward. On a federal officer's approaching the house with the purpose, she surmised, of paying for accommodation of his detachment, she ordered kindling lighted in the fireplace "to take the chill off the morning." Standing, thanking her caller for his courtesy, she dropped the greenbacks into the flame behind her and was wiping her fingers on her handkerchief as she accompanied him to the door.[14]

Frederic was one who could accept Yankee money with a noncommittal nod and use it, if need be, to meet any indemnity a Freedman's Bureau might affix as owing to Blacks who had remained in his service two years, six-and-a-half months after Lincoln himself had read the Proclamation on January 1, 1863. Walnut Creek Blacks would settle their claims without need for military reinforcement. They would then rent land and provide labor, the landlord to furnish teams and tools, the crops to be shared half and half, until they could become landowners in their own right. Black women would continue to nurse the sick and tend ladies having babies as they had in the past, only now they must be paid for their time and trouble—though not

---

13. This incident quoted by Jody as part of family lore.

14. Others, like Frederic, were not above making every edge cut they could. Eugene, collecting lore and reminiscences, heard from Dan Matthews of Govalle (an Austin suburb) that his family secured potato slips (for planting) from potato peelings cast on the Federal army's garbage dump. Letter, J. L. Alford to Eugene, November 11, 1939: "When the northern army came to Texas (1862-1865) and camped at Fiskville they bought a lot of corn from grandfather." Mrs. Rose: note from Fannie E. Ratchford.

Govalle: the writer has been told that this name for a fertile truck gardening area came from Swedish settlers who, when the area was some distance out of town, were given to saying, as sundown approached: "Well, I must go valle now!"

necessarily in cash. Cash money was scarce with everyone. A bit of fine sewing or a couple of squares for a fancy pieced quilt would be acceptable in return for a family wash day and a share of the meat for help given at the hog-killing.[15]

The thrifty would save to buy homesteads; it was ownership—the ownership of land—that made a person not only proud but able to send sons and daughters off to one of the new schools for Black folk.[16] None saw lifting of the pall of slavery as a signal for extinction of accustomed social levels, whatever General Sheridan or northern abolitionists or a federal Congress might think. That which did have to be done immediately, however, was the choosing of last names. General preference was for the name of the first plantation master recalled by one's oldest ancestor, though some took this upward step by claiming the name of those of whose paternity they were certain—as did Judge Hemphill's daughter, Theodora—and others who knew themselves to be among Sam Houston's many descendants.[17]

A placard at the Texas port of Indianola, while secession was being whipped up, had proclaimed COTTON IS KING! Return to a semblance of normalcy found Frederic with his mind more than ever on cotton. The South was broke and it was beginning to be said that, war or no war, planting over and over in the same rows had so exhausted the soil as to play havoc with the southern economy. His grandson, John Lafayette Alford, would remember a conversation heard in his childhood when a neighbor, lauding the possibilities of prairie production, once the sod was broken and plowed under, said,

---

15. Jody told of piecing two quilt squares weekly to get the family wash done.

16. According to George Elroy of Texas Southern University's Journalism Department, writing for the *Houston Post,* February 9, 1978, twenty-seven colleges and universities had been established for Black students by 1870, the majority of them in the South.

17. Mrs. Eliza Norwood Shanks and others have told the writer this was how names were chosen. It compares with the naming of Swedish settlers, most of whom, according to Miss Joanna Almquest, had no surnames when they came, hence the falling back on Swen's son, Ander's son, and the like.

There were young Houstons among Texas Southern University students when the writer taught there. It was an observation of the writer's father that Sam Houston had left his children, "black, white and brown, all up and down the Brazos and Colorado rivers"—which was no different, except for its geographical extent, from what one hears of other gentlemen of consequence during the same era. The "Victorian period" in relation to men was the exact opposite, morally speaking, from its relation to women.

"This black land won't wear out," and his grandfather's response, "No, it won't wash."[18] Had they asked the Indians, the Indians would have told them better.

Frederic had lost a certain amount of property value due to emancipation, but, counting the 99 acres left in Bastrop County, his near two thousand acres gave him a feel of prosperity and an impetus to start life over again in accord with new post-bellum rules. The storms of middle age were past; Indian summer was yet to come. All around him Black folk, instead of rushing up North to the fabled Land of Liberty, were settling to respectability—their medium of exchange, the labor of their hands.[19] Without them, cotton was unthinkable because only their complexions would withstand the sun: cotton—with its five fleecy locks exactly fitting the five fingers. He had lost his elder sons in a war he had hoped to avoid, but John and Zackie had been below the age for conscription. Zackie, sixteen, was as yet interested mainly in hunting and fishing. He would leave early of a morning on Frederic's riding mule, "Ole Yaller," often with ten-year-old John Lafayette mounted behind him; it was his pride and joy to return with a full game bag for the family table.[20] In height, he topped his brother, John, who was twenty, by more than a head. It was John, whom grown-ups in the family called "Bud" and the children called "Uncle Bud," whose thought turned to what was beginning to be known as "The Field."

Harriet was from South Carolina; Frederic (whom his children called "Pappy"—they called Harriet "Mother"), of French descent, was from North Carolina and enough of his European background remained with him for land to be synonymous in his mind with lineage. In that regard, he was not ill-pleased to have Harriet's Bible show (besides his eldest daughter's three children and George's three) that Mary Catherine and Will Alford had four sons: Thomas Frederic,

---

18. Reminiscence in letter from J. L. Alford to Eugene, November 11th, 1939.

19. A population check (*Texas Almanac*) of census rolls for the years since Emancipation shows no exodus of Black people from Texas. Nor has such a movement corresponded to the shift of cotton planting from Old Texas (Texas below the Hill Line) to West Texas plains and a climate less hospitable, socially and climatically, than that below the Balcones. The movement of Texas Black people has been rural to urban, as with all Texas ethnics, giving Houston and Dallas particularly growth in accord with the loss of farm population.

20. Ole Yaller: at least from Chaucer's time, a riding mule was considered the safest and most comfortable mount for ladies, priests and elderly gentlemen.

nine; William, seven; Baldy Samuel, six; Charles Ferdinand, two; and a daughter, Jennie Louise, four. Another child would come, toward fall. Their house was within shouting distance of his own.

All his daughters were married except twelve-year-old Julie, who still addressed her sisters in the fashion picked up in babyhood from Blacks (whose pet she had been) as "Miss Em'ly, Miss Mary, Miss Sallie, Miss Jenny"—though she distinguished Amanda Bradford as "Manch," for what reason no one knew, unless in infancy she had confused "Amanda" with "Comanche." Sallie, married to Lt. William Leroy Giles in '66, had two little dark-haired girls and was "expecting" again. They had built a house where the old Cameron Road ran through The Field on the west, and had added a separate one-room dwelling for Dock's parents, grown too aged for farming.[21] Sallie, who had never cooked a meal in her life until she married, was becoming proficient in handling her iron cookpots at the open fireplace.[22]

Jennie, married a year later than Sallie—to Lt. James Corry Maxwell, late of Hood's Brigade, was living in what was known as the "yellow house" (for its paint) where Maxwell land touched Frederic's—the new land to which Mary Corry Maxwell, "Jim's" mother, brought with her, in her trunk, papers dated back in the 1700s showing Corry land grants in the Old Dominion.[23] So far, the young Maxwells were childless. Amanda and Dan Payton and their two little boys, John Wesley and Ephraim Frederic (who were double cousins to George and Lucy's children), were still living with Dan's mother, Mrs. Payton-Forehand, but would soon move into a place they were buying at the foot of The Mountains. Lucy was married to E. C. Moore, a prosperous Bastrop farmer.[24]

Harriet Emily, whose young Alfords were double cousins to Mary Catherine's youngsters, was also remarried. She was living in The Field with her new husband whom nobody knew anything about (a German

---

21. The Memoirs mention such a room, which, after the old folks were gone, was used as Eugene's room.

22. The Memoirs quote Sallie as making this statement.

23. Jane gave these papers to Jim Corry after her husband's death and they now are said to hang, framed, in the John Corry home, Enid, Oklahoma.

JANE (JENNIE) and James Corry Maxwell.

immigrant listed in the Census as a wagoner and as several years younger than his wife), whose name was William Alfred Waldemore.

Looking over the family record of marriages, births and deaths—and then over the acres greening with cotton under management of his son John and his several sons-in-law, Frederic concluded that the land would take care of them all.

---

24. Edward Cobb Moore sold his store and farm at Fort Smith, Arkansas, and removed his family (three grown daughters and 17 year old son, Taylor, to the Prairie Lea Community in Texas, in 1863—taking along those Blacks who remained with him after the Emancipation Proclamation. His wife (Agnes Twitty) died en route. Reference: Victor Moore, Taylor Moore's grandson, El Paso, Texas.

Good Man Gone Wrong

From ghaisties and ghoulies and lang-leggedly beasties and things that go bump in the night, good Lord deliver us!
AMEN.
From witches and warlocks and things that run from the light good Lord deliver us!!
AMEN.
From battle and murder and sudden death, good Lord deliver us!
AMEN. AMEN. AMEN.

—Old Scottish Prayer

The Iron Clad Oath, together with the army of occupation, was withdrawn from Texas March 30th, 1870. Edmund J. Davis (previously Judge Davis and late a brigadier general in the Army of the USA) was left as governor to see what he could do with a wrecked state and a spiritually unreconstructed people. He had been in Texas since his arrival from Florida in 1838, except for the war years; nevertheless, in the bitterness over a militarily-supervised vote of Unionists and those Freedmen who had gone with them to the polls—all lumped by the disenfranchised opposition as "Black Republicans"—he was referred to as "the carpet bag governor," an epithet contemptuously coined for incoming opportunists carrying their worldly goods slung over the shoulder in bags made of Aubusson carpeting.[1]

A legislature chosen by the same constituency authorized organization of a state police to abet the Ranger force (embryo when Texas was a Mexican colony) in bringing order out of chaos. Davis invited Confederate veterans to take commissions over Black recruits.[2]

1. "Scapegoat" was the word for a turncoat; neither that nor "carpet bagger" fitted Davis, but "carpet bagger" had to do.

2. The Ranger force is under direct command of the governor of Texas. Stephen F. Austin, as head of government for Mexican Department of the

Among those who did not spurn his invitation was Lt. William Leroy Giles. Dock's view was that money was money and must come from some more dependable source than a few bales of cotton still in the soil. His commission is dated August 9, 1870. Moreover, if ever the country needed his service, it was now, what with the Hill Country forts not yet reactivated to keep the Comanches in check, and so much lawlessness afoot that there were even murmurs of "a good man gone wrong" over Gilliland way who had turned to preying on his neighbors.

But despite encroachments from without and lawlessness within, by the time the telegraph line reached Austin, and, on January 4th, 1872, the Houston and Texas Central Railroad had connected the capital city and the port of Galveston, something new was in the air: with more railroads pointing Southwest, a few daring men, rather than bemoan the plantocracy's vanished dream of empire, fixed their minds on the need for beef in northern markets. These men were organizing "outfits" to go on the Mustang Prairie in the Brush Country below San Antonio, cut a few hundred head out of the vast herds grown from cattle abandoned with the Spanish missions and take them on the hoof "to meet the railroads."[3] So quickly did this idea catch on, it soon was the hope of every adolescent and post-adolescent to join such an outfit and make at least one trip "up the trail." Meanwhile, Judge Sneed and his friends who had loped off to Mexico in the hysteria wrought by collapse of the Confederacy were back in their offices, doing what they could toward bringing legal confusion to an end.

In the big house without the columns (where she had been sent

---

Brazos, had at his command a force of vigilantes to which Jared Groce contributed, when emergency arose, a company of thirty young Blacks (reference: Mrs. Berlet's ms.) trained in military tactics by himself.

First official record of the Rangers appears in the *Journal of the Constitution,* compiled, 1852, by James W. Robinson, erstwhile provisional lieutenant-governor of the Republic of Texas: "We have (November 1, 1835) authorized raising of twenty-five volunteer rangers, to range between Colorado and Brazos on the frontier settlements; twenty-five between Brazos & Trinity Rivers, and five to ten—the east of the Trinity promising each volunteer one dollar and twenty-five cents a day."

3. Frederic's signature appears with fifty-eight others in the *Austin Statesman,* October 24, 1873, petitioning the county court for an election authorizing $250,000 in bonds to be donated to the International and Great Northern Railroad to help bring its trains into Austin by May 1, 1894.

by stagecoach and alone after her mother's death, when she was five years old) sixteen-year-old Jody Butler, Mirinda's niece and the judge's ward, was married to Frederic's son, John, on January 16th, 1873. John would take her home to his parents' house on Walnut where only Zackie and Julie were yet unmarried.[4]

Next morning after the wedding little Belle Alford, catching a glimpse of the bride before she was fully attired, would forever remember the silver garters that captivated her fancy rather than the red-gold hair or the dove-grey "second-day" dress with blue ribbons. What Jody would ever remember was the sight of John's fourteen white ruffled and embroidered shirts (with one colored one) which she, who never before had been responsible for anything, would be responsible for seeing were kept properly laundered!

Some months after Jody joined the family circle, Will Alford died, leaving Mary Catherine to manage their property and finish bringing up their seven children—by then inclusive of Julie's namesake, aged four, and Mollie, two. In January of the next year, her father would help her sell for $1800 cash and a $200 note at 10% interest the 274 acres bought by her husband in June of 1870 from what had been Elizabeth Boyce's place on Gilliland; thus her home would remain near him and Harriet.[5]

This was also the year a group petitioned for a change in the Cameron Road to Austin—a change from the Jourdan Crossing on Walnut to a crossing a couple of miles south, leaving the old road as a loop past Sallie's house to Dessau and on, following its original path, to join the new route below C. C. Baker's gin and the Gilliland Creek bridge, above which was William Gregg's store.[6] From Gregg's Corner, a right angle turn-off led to Wheeler's Store, before its juncture with another main road going down the Colorado to Webberville and Bastrop.[7]

---

4. According to Jody, they lived with the "old people" two years, during which time she was always made to feel welcome in their home.

5. Travis County Records, Deed Book T, p. 320 and Deed Book Z, p. 353.

6. The petition (copy from Mrs. Florence Johnston Jarmon, Austin) was signed by Thomas J. Jarmon, William Neans, A. B. Townsend, William Custard, S. L. Jarmon, L. B. Giles, J. C. Maxwell, Thomas Maxwell, C. A. Maxwell. Did Frederic oppose the change? One wonders, since his name is not on the petition.

7. With arrival of the Houston and Texas Central Railroad in 1871, "Wheelter's Store" became the town of Manor, named for J. B. Manor, who gave the land for the townsite.

Not long after the road was changed, a widowed Virginian sold his harness and saddle shop in the Old Dominion's town of New Rochelle and came to Texas to his brother, James P. Sprinkle, at Hornsby's Bend. Frederick Erasmus Sprinkle brought with him his widowed daughter, Cremona Sprinkle Barr, her three young sons: Robert Hayward, William Braxton, George W., and their sister, Ida.[8] Encouraged by proximity of plantations and the travel along Cameron Road's new route, he built a commissary just east of its Walnut Creek crossing.

Late in January, word came up from Bellville that John Atkinson was dead. He had been in his eighty-second year. Since April of 1857 he had been married to Mrs. Mozelle Wilkins.

It was the dawn of a lovely Indian summer day in 1875 when Frederic started a train of cotton—three wagons drawn by mules, loaded with three bales each—over the old road to Austin. One of the drivers was Jack Black, brother-in-law of William Shanks, who was Amy Black's husband. Others were his brothers, Sam and Bill Black. The Black family had "come to Texas with the Maxwells" and as Freedmen were also farming in The Field. Jack would remember that October morning so long as he lived, which would be long after the others were "laid away." The wagons had been loaded the night before; the mules were hitched and ready when Frederic came out in his high hat to follow the train in his one-horse phaeton. He would sell the cotton (according to old account books) for around $60 a bale, each bale weighing more or less five hundred pounds. It had been ginned over at Maxwell's "yellow house" where Jack Black worked during the ginning season and where Jennie herself sometimes fed the raw cotton into the horse-powered "stand" where the lint was separated from the seed. They could, in Jack's words, "turn out ten bales between kin an' k'ant"—that is, in the working hours of "daylight to dark."[9]

Back at the house, Sallie was visiting for a few days with her mother and Julie. She had with her her two youngest—Flora, who was four, and Maggie Belle, two. She had not been feeling well and it did her no good that Flora was fractious, going into tantrums when she did not succeed in simple household tasks like Alice and Laura, or when she could not identify words in books as they did. Maggie, though dif-

---

8. J. P. Sprinkle, Erasmus' brother, married Margaret Hornsby, April 13, 1859. Travis County marriage records.

9. See Appendix C.

ferent, was also a problem. Almost too docile, when she tried to make her wants known she could manage only incomprehensible gutturals as if her tongue were too big for her mouth. Small as the older girls were, they had stayed at home to help their father and give their mother a rest.

John and Jody had recently moved into a house built for them over in The Field, so, except for Mary Alford's ten-year-old Baldy, there would be nobody but womenfolk on the place until Frederic's return, Zackie having ridden off with his friend, John BCoat, probably to be gone until late in the evening. (He and BCoat were keeping "bachelors' hall" in a small structure apart from the main dwelling.)[10] "Going up the trail" with cattle outfits had made Zackie something of a man of the world. Thanks to the route the trail drivers took, traffic on the road past the house was more frequent. It seemed nothing unusual to the women for a party of riders to unsaddle at the Crossing and noon in the shade whence they could be heard joshing each other over a card game. To tell of the next twenty-four hours was to be the greatest of all family folk tales:

> No strangers were in sight when the cotton wagons came home empty at dusk.
>
> Frederic was weary. Supper over, he spread a pallet on the gallery and in minutes was asleep. Sallie put her daughters to bed in the east room, lying down with them to get them quiet. Harriet took up her sewing in the west room where Julie and Baldy joined her after cleaning the kitchen. With little to talk about, the evening seemed preternaturally silent until a screech owl's startled shriek—nothing like its customary quaver—split the dark. And Frederic thundered, "What do you want?"
>
> Baldy, thinking the owl might have added to a nightmare his grandfather was having, dashed through the hall to wake him, but halted at the open door. Outside, a shadowy figure was demanding, "Get up, old man! We want your money—that's what we want!"
>
> Seconds later, the boy was back in the room, his grandfather following him, herded by two men with guns drawn and hats pulled low to the bandannas masking their faces. Although one dangled a *moral* (a Mexican nosebag for feeding horses when on the trail), they were not Mexicans—they were Anglos. The guns motioned Frederic to his small steel safe against the far wall. At each window (there were two, with the fireplace between) other

---

10. A separate "quarter," so to speak, for young men in the family was a usual part of plantation arrangements.

masked faces peered in. Frederic knelt before the safe and the men moved close, clicking guns at his head. His hands trembled so that he fumbled the combination, then, as a gun prodded him from behind, a crash in the east room made the marauders whirl to spring across the hall and into that room where Sallie, lifting from her pillows, reassured them politely, "Be not alarmed, gentlemen, it was just my little child falling out of bed."

Again in the other room, the foremost man, seizing Julie as she was about to slip over the windowsill—the eyes outside had disappeared at sound of the crash—pulled her back, telling her as he patted her curls, "Why we wouldn't harm a hair of this pretty head!" Then, turning to Frederic, he declared if the safe were not opened "instanter!" he would blow his brains out.

Again Frederic's hands were unsteady while Harriet, kneeling by him, adjured the Almighty to expend his wrath on those who would rob an old man of his hard-earned gains, and Baldy, dancing in horror on his other side, begged, "Oh, hush, please hush, grandmother, they may kill you, too!"

The heavy door swung clear at last. The man with the *moral* reversed his gun, smashed the safe's little wooden drawers with the butt, emptied the contents into the bag and in moments only the ring of horses' hoofs on the hard ground, vanishing in the distance, remained of the raid, except for a taunting shout flung back that every house in The Field would be "cleaned out" before the night was over!

When silence was absolute, Julie and Baldy crept forth to waken William Shanks whose house was near. For one wild moment as they moved through the dark, clinging to each other, they thought there was pursuit but it was only Julie's heart thudding in her terrified bosom.

Within the hour, Black men had alerted every family in The Field and the news was spreading up and down Walnut and beyond. At John's and Jody's, John was readying his guns while Jody's sister Fannie (Fannie Butler, resident with them at the time) was sewing greenbacks he had taken from his wallet into a coat sleeve to hang innocently from the wall peg. Jody, fearful for her life if so much as a breeze rustled the grasses, cowered beside her baby boy, Eddie, wailing thinly as if the molesters were already upon them, "Give it to them! Give it to them and let them *go-o-o*!" distracting him till his hands shook like his father's. He urged Fannie, his eyes on the bottle and spoon beside the bed, "Give her some more of that nervous medicine! Give her some more of that nervous medicine!" while Fannie, who was older than her sister and knew her better than he did, went right on with her task.

> Zackie and John BCoat, coming home late without running on any of the Black couriers, were astounded to find saddle horses tied round the house and more astounded by noises coming from within where Old Man Rowzee was brandishing a razor, imperiling those near him and deafening everyone with roars of what was to happen should he come on the robbers in person! (In the jargon of the day, "Old Man" was a title respectfully accorded the head of an established family or any ancestral patriarch.)
>
> By morning, excitement was boiled down to knowledge: the robbers' parting alarm had been but malicious braggadocio—no other place had been visited. Frederic had banked his cotton money before he left Austin—but the safe had held several hundred dollars of Mary's and a fifty dollar bill of Zackie's. This bill the robbers dropped unnoticed and Harriet had not been so engrossed with addressing her Maker that she failed to sweep her skirt over it as it touched the floor. Conclusion was that, without doubt, the dastardly deed had been the work of Neal Cain's gang from over on Gilliland, and that the "big man" Baldy had seen holding the horses ready for the getaway was none other than Neal himself.

The episode was but one of many in the country's general disorder and a further step in a career of crime undertaken by the eldest son of a landholder whose daughter, Mary, was Jane's best friend.[12] Jody's husband, thinking things over after his wife had been soothed and his guns put away, recounted for her and Fannie the times when, as a boy, he had hunted, fished, even camped with Neal, nine years his senior and therefore a flatteringly pleasant traveling companion.

At a later date, Dock Giles, reminiscing, would say one might not even turn one's milk cows out to graze because an older, hardened Neal kept the pastures on Walnut and Gilliland stripped, smudging brands and melding stolen cattle into trail herds being driven north. "Why, you might hear your cows bawling in his pens for their calves, drive your calves up to the fence and have the cows trying to get to them through the rails, but it would have been worth your life to start taking the rails down to let the cows out."

In the opinion of the country-side, Neal's wickedness went still further. Two women, Jody told the writer, were murdered on Gilliland, one of them his sister; when found, the sister's baby was on the floor with the dead, paddling in her mother's blood. It was surmised that members of Neal's gang did the killing "because the women knew too much and might have exposed what they knew."[11]

In the fall of 1871 Davis had been voted out of office by reenfranchised Democrats. He declared the election illegal and was said to have barricaded himself in the capitol basement with some of his "Black police" while he wired President Ulysses Simpson Grant for reinforcements. The president-general wired back that "the people have spoken, their wishes must be respected"; and the Republican Party, though it might linger awhile, like a breathing corpse hitched to an oxygen cylinder, was done for in Texas. It remained to be seen what post-bellum Democrats would do. The first evidence was the Constitution of 1876. (It is still in effect.) "All Men Have Equal Rights" was the initial line of Article I, Section III. Despite this constitutional assertion, smouldering Red-Hots, who had been sent to the legislature wherever plantocrats could muster a majority, set to work to achieve what northern abolitionists had feared would come about—covert oppression of Freedmen. It would be done in legal accord with the North's ideal of human equality—by bringing over Europe's "huddled masses" in such numbers, such poverty and with such ambition to achieve that "upstart Blacks," unable to meet the competition of wage-paying jobs, would be forever compelled to drag cotton sacks and scrub floors, meanwhile, knocking on back doors, begging

---

11. The story of the "hold-up" was so often told and retold in the writer's hearing it is worthwhile to cite but two letters of reminiscence in the Giles Papers—one from Baldy S. Alford, New Orleans, December 27, 1939; the other from John Lafayette Alford, McDade, Texas, November 11, 1939.

The writer has found no news reference to the murder in the Cain family but presumes the baby to have been the girl-child, Belle, reared in the otherwise childless home of Neal's younger brother, S.M. (Sugar) Cain and his teacher-wife—the "Aunt Hattie" beloved by the William Boyce children as well as the Aaron Boyce children whose mother was Jane's friend, Mary Cain. These were the grandchildren of Elizabeth Boyce who, like Diana Gilliland and many other Texas women, had her own cattle brand and is listed independently in the Texas Census as "a farmer."

Cornealius and Sugar Cain (listed in the 1860 Census as "Sugar Molasses Cain"), sons of Joel Y. and Philadelphis Cain, were nephews of Sugara Macklemore Cain—from an old family established near Hot Springs, Arkansas. Sugar and Joel Y. seem to have been business partners with a spread of land on Gilliland. After Sugara's death in 1869, his wife, Susannah, married W.D. Edwards, who, Eugene says, had one of the gins in the area. Some of this information is from a glass slide picked up in an old Austin building and placed in the Austin and Travis County Collection: apparently part of a contract, preceding Susannah's second marriage, in regard to a final property division with Joel's heirs.

for kitchen hand-outs and to be allowed to sleep in their old quarters as the only escape from exposure to the elements and starvation.

Bare mention of this policy roused Frederic to fury. "Bring these Dutch over here," he stormed, ("Dutch" was indiscriminately applied, in popular parlance, to any non-English-speaking North or Central European) and you'll be robbing your children of their birthright! *They'll* be the ones to own the land before you know it!" Neighbors, amused to think of Harriet's father and his language problem, would tease, "But Mr. Jourdan, how can you cuss "the Dutch" the way you do when you married a Dutch woman?" And he would respond in a roar, "I may have married a Dutch woman, but I didn't marry a *damned* Dutch woman, and these are the *damned Dutch* I'm talking about!"[12]

At times he would be more explicit: he would say, "There are three kinds of Dutch—high Dutch, low Dutch and *damned* Dutch." To his listeners this only meant that, once a natal language was dropped in favor of English, the folk under consideration would turn out to be a mixture of "nice people, common people and trash."

A land dealer to his fingertips, he had his own idea of patriating Blacks. A year after the new constitution went into effect, he sold William Shanks 40 acres in The Field for $690 cash and a $647 promissory note due in five years at 12% interest.[13] It was the beginning of a series of sales to Freedmen whom he had known well as frontiersmen while they were still nominally enslaved.[14] His feeling about "the Dutch" was soon to be put to the test on his own doorstep. Harriet Emily's husband, the young immigrant wagoner (listed in the 1870 census as from Germany and in the 1880 Census as from Switzerland) disagreed with John over management of The Field. Asserting his European idea of a superior position because of his longer years and his marriage to Frederic's eldest child, he drew back, threatening a blow to the head with a singletree.[15]

John, having gone to The Field unarmed, dismounted at Sallie's house, nearer at the moment than his own, asking for a gun. Dock, the level-headed, judicious counselor, calmed him, promising the family would confirm him as his father's deputy. Upshot of the quarrel was that Waldemore took Harriet Emily, her son, John

12. Frederic's ethnic prejudices are still being quoted, with smiles, by his descendants.

13. Travis County Records, Deed Book 55, p. 166.

14. See fn. 4, Chapter Four, "The Distant Tocsin."

15. William and Amy Shanks.

Lafayette, and her younger daughter, Belle, down river and across to Evergreen on the Yegua where original land grant families were already, as Frederic had predicted, selling out to incoming Europeans. Harriet Emily's namesake daughter (called Dude) remained on Walnut, married to Virgil Edrington, engineer at the Maxwell gin, who had been a compatriot of his uncles-in-law, Jim Maxwell and Dock Giles, through the Civil War. Sadly, their honeymoon was scarcely over when a fractious horse kicked the young wife in the chest. She lingered for a time suffering from an injury never fully diagnosed, but dying before the year was out.[16]

Waldemore's removal was no happy end to the fuss between him and John. He had never been kind to his steplings and would continue to be unkind, holding even little Belle out of school for weeks at a time to drop corn or do other field tasks. The Jourdans, even-tempered for the most part, especially where children were concerned, had they hated anyone, would have hated him, and all the more for his taking his part of the family into what they looked on as a kind of exile.

About the time of Harriet Emily's departure, Mary Catherine lost her third son, William. His death was sudden. Because the children had been in the orchard the day before, eating green fruit, some thought his trouble was cholera morbus—a name for acute indigestion; others, that it may have been a heart attack.[17] A third possibility entered no one's mind, for acute appendicitis was then unheard of.

Although it had nothing to do with William's death, horses played a considerable part in the overall family's ill luck. Harriet's maternal great-grandfather was the Count Zimmerman (Harriet thought "Zimmerman" was the right name). The count had two daughters; the younger (Harriet's grandmother) ran away to America with a Protestant (Lutheran) minister, Charles Beneke. When their father died the elder daughter inherited the title and estate and came to visit her sister—presumably in South Carolina—where she was thrown from a horse and killed, making Pastor Beneke's wife the heiress. Madam Beneke said she had crossed the Atlantic Ocean once and would not cross it again for all the lands and titles in Europe.

---

16. Data on Dude's death from Emily Swartz Rutland, Kingsville, Texas, who paints the earthy genre of the country in earth colors. Jim Maxwell had been among the country's earliest ginners to change from horse power to a steam engine.

17. Data from Minnah Lea Davis Mason (Mrs. Clabe Mason), Mary Catherine's granddaughter, Durant, Oklahoma.

Since it was then the fashion to declare oneself an all-out American (to whom titles and entailed estates were constitutionally forbidden) old John Bachman of Orangeburg, whose first wife had been Mary Catherine Barbara, Madame Beneke's daughter, cared so little for ancestral glory that after her death he let his children play with the pretty seals on the estate papers till the record was destroyed—at least that is how the story was remembered. It is not even certain if the estate was in Germany or Alsace.[18]

Up Walnut from the Jourdans lived Sinnet Musset, a horse trader, racer, and gambler who had married Judge Sneed's second daughter, Mary Ann, back in Arkansas when she was thirteen—so Jody said. Musset brought her out to Texas in 1847 where she died in 1849 at the age of twenty-six, leaving a new-born son and a daughter. The daughter, Polly, married Eph Mayfield and it was to her house that Jody ran from a German family where the Judge and Mirinda parked her when they fled to Mexico. Musset, as Eugene recounted to the writer, introduced an extraordinary type of horse—the Steel Dust horse—to the "Jourdan flat," a quarter mile of pasturage that stretched south of the house toward the creek and was perfect for quarter-horse races.[19]

---

18. This story is told in all branches of the John Bachman family; Harriet, only twelve when Mary Catherine Barbara died, may be excused for not knowing too many of the details.

19. Musset: listed in the 1950 U. S. Census as "Tennet Mussell," with a six-year old daughter and two-year-old son, married Mrs. Susan Robinson, widow of Alfred Robinson, in 1866, Travis County marriage record. Evidently he was a handy man with a gun; Dobie, in *Coronado's Children,* "The Lost San Saba Mine," describes his lightning dispatch of an enemy at a San Saba bar before the enemy could draw down on him. Steel Dust, the sire of the strain brought to Collin County, Texas, in the 40's, wasn't much to look at, according to his biographer, Wayne Gard. It was his speed that counted. His mouse color plus his stamina, one must assume, gave him his name. A quarter horse, he was insignificant as to size, especially when lined up with the county's big champion, Monmouth, on Sunday, September 5th, 1856, at McKinney (population normally 209, increased to more than 2,000 on the day of the race), where he broke every betting man in the North Texas-Red River area from Oklahoma Indians to soldiers and settlers at Fort Worth and Dallas. Ref: *The Examiner,* McKinney, Texas, December 6, 1856. Throckmorton, of the Immortal Eight (see fn. 9, Chapter Four, "The Distant Tocsin") started the race in frontier fashion with a wave of his wide black hat. Throckmorton was then on his way to the governorship of Texas in 1865; however, his political career was cut short in a year by arrival of General Sheridan with the army of occupation.

What this may have had to do with Frederic's disposal of his last Bastrop property, 99 acres, is a surmise. The fact remains that on February 16, 1878, he gave John H. Pope of Travis County a quit claim to this acreage for the token sum of five dollars.[20] And there is a family story that on some occasion, "he traded his land for a fine horse and the very next day the horse lay down and died!"[21] Could he have been going in for raising Steel Dust horses and such a horse have made up the difference between five dollars for 99 acres and the previous price of six dollars an acre for the rest of his Bastrop land? If so, this was one horse Neal Cain failed to brand.

The new Constitution dealt extensively with the organization of a court system. By the spring of 1879, Neal, tracked down by the Rangers, was under indictment but was out on bail and it was rumored that he was about to turn state's evidence to save himself from penal servitude.[22] He was then living with his family at Taylor in Williamson County, thirty miles out of Austin. Taylor was on the International and Great Northern Railway that had run its first train into the capital city on Christmas Day, 1876. On Monday, April 22nd, 1879, the noon train brought word of Cain's assassination. He and his wife, a small woman, had gone peacefully to sleep Sunday evening with her between him and the window whence the breeze came. Naturally a

---

Lonrigg, in *A History of Horse Racing,* p. 221, says ". . . there were hundreds of meetings at county seats and railheads; hundreds more, especially in Texas, unofficial and unrecorded at cow-towns and behind houses . . ."

Goodall, in *Horses of the World,* derives the term "quarter horse" from a mixed ancestry—English thoroughbreds with mustang cow ponies, but Texans incline to derive it from the informality of the early race courses: turnrows in the fields and pasture flats like Frederic's that were apt to be no more than a quarter mile long. Interviews (1977) with Tex Rogers and David Gaines, editors of *Horseman,* Houston's Magazine of Western Riding.

20. Bastrop County Records, Deed Book Y, p. 157.

21. It would seem that this horse left his brand on the psychology of Jourdan descendants for they have bought no blooded animals, always sidestepping temptations with the excuse, "It might just lie down and die!" Major Leo G. Staly, DVM, USA, interview March 2, 1977, says a horse might have unexpectedly dropped dead of anthrax, a virulent disease among Texas livestock in the '70s.

22. Lengthy reports of Rangers' attempts to take Cain in custody appear in the *Austin Daily Democratic Statesman,* October 28 and December 1, 1876. Davis' state police was disbanded in 1873. J. Frank Dobie, in *Coronado's Children,* places John Ringo, who was with Cain on November 26, 1876, over in Mexico rustling cattle, after Cain was wiped out.

HARRIET EMILY III and Virgil Edrington.

large man, his paunch had increased with the years (he was in his middle thirties) and was a plain target in the slanting moonlight. The bullet cleared his wife to leave him in agony for which there was neither antibiotic nor anesthetic. He died at dawn as he had lived—desperately.

This story was given scant inches next day in the *Austin Daily Democratic Statesman.* Legendary details more befitting a man of such character were handed down to the writer by her father who told her it was said at the time of the assassination that gangrene had set in, swelling the body till it would go in no casket so a box had to be hastily carpentered to accommodate it. Also, that only Austin gamblers and saloon keepers on horseback were in the funeral cortege when a tremendous storm broke, loosing thunder-bumpers that threatened to jar the boards apart and lightning that danced on the makeshift coffin until every rider turned back to town, leaving the corpse to the fury of the elements. Thus told, the story is embellished with the ancient mystique of the Devil's claiming his own before an evil spirit's mortal remains reach sanctuary in consecrated earth.[23]

Something, hardly an epitaph, appeared on April 24th in the Austin paper:

> Since the exit of Ed Tracy and the killing of Neal Cain and the Lemons, the stock business is looking up in this corner of creation. The grand jury in a few days will thin out a few more of them.

---

23. The writer, unable to locate Cain's grave or learn toward what cemetery the cortege was headed, is sorely tempted to accept the ancient, animistic surmise that "the Devil done tooken his soul and flew off with it!"—an idea based on the concept that the Evil One must claim his own before the mortal remains reach sanctuary in consecrated earth, for so long as a fragment remains, the soul may return at will to its worldly habitat. The same notion was put forward by Blacks when Madame Delphine Lalaurie, who tortured her slaves, died in New Orleans; also by Moors in Spain when Queen Isabel's body was carried down the Peninsula to be interred in Granada—that her spirit might forever be present to hold the "Infidels" subject to Christian sway.

The Hunter's Moon

From gold to grey
Our mild sweet day
Of Indian summer fades too soon.
—John Greenleaf Whittier

The decade of the '80s appeared to begin auspiciously for Frederic and Harriet and their extended family. Julie, like many young ladies of twenty or so, was teaching a home school, imparting to neighbor's children and her nieces and nephews the rudiments of learning known as the "Three Rs"—for which she received some such sum as two dollars a term per pupil.[1]

For the past ten years, Amanda and her husband had lived on their stretch of land along the road where the Mountains crept down to the prairie, a couple of hours (in a buggy) west of Walnut.[2] They then had four sons: John Wesley, Ephraim Frederic, Andrew Jackson, James W., and a small daughter, Eula Lee. They expected another child in late spring. (That one would be his father's namesake, Daniel.)

Sallie had adjusted her life to her "little afflicted one," as she customarily referred to trantrumish Flora (whose portraits and reported behavior indicate an epileptic mongoloid); to Maggie Belle, almost too docile (and certainly a cretin) who, though no more able to learn than Flora, only cried quietly instead of going into tantrums when her feelings were hurt; and to loss of a beautiful little boy, Willie (William Leroy), bright beyond his two years, eight months, when he died on November 27th, 1878. Fourteen months after they buried Willie beside the fresh grave of his Grandfather Giles in the Fisksville Cemetery, and two days after Saint Valentine's, 1880, she and Dock welcomed a second son, Eugene Victor, who gave promise

1. Sallie's letter, note 10.
2. Abstract provided by Pauline Payton Johnson.

equal to the lost Willie.[3] The baby, Eugene, was twin cousin to John and Jody's second son, John Fields, born on Washington's birthday. Though any new baby meant a host of relatives on horseback, in buggies, in spring wagons, or whatever—come to view and appraise the newcomer—it was the wedding of Mary Catherine Alford on September 4th, to Frederick Erasmus Sprinkle, that brought the family as a whole its greatest happiness since post-war marriages back in the '60s. Never was a stepfather welcomed with more glad acclaim than "Uncle Sprinkle"—and that without thought of his being but two years younger than Frederic.[4]

Nine days from the wedding, little Daniel died. He, like Willie, was buried in Fiskville cemetery's white chalky soil—all alone in the northeast corner. No one thought of this sad little death as presage of an avalanche of sorrows. Rather, thought was that Amanda, with her growing brood, was fortunate to have suffered no such ill fortune as had befallen Sallie with Flora and Maggie Belle. Though not spoken aloud, everyone knew of the whispers that Dock Giles, distinguished and respected neighbor that he was, must have been a bit in his cups when those two "off" girls were conceived.[5]

If there was a further problem to consider, it was Julie's courting with Osceola Claiborn Cato—a slim dark young man, a cowboy par excellance, who, having left home at fourteen to "trail the cattle up north," had been, because of his skill with the herds, advanced at eighteen to management of a Central Texas ranch.[6] On September 17, 1879 (two and a half months short of his majority), he bought five head from Julie's Uncle Dock for the sum of $27.50 to start a herd of his own.[7] Considering that a cowboy's wage while he furnished his own horse and saddle might be a dollar a day, this was no deal to sneeze at.[8] Nevertheless, however smart and ambitious Ocie was, he possessed not one acre to tie to—and cowboys are notorious drifters.

---

3. Willie was often mentioned in the writer's hearing, particularly by Jody, as an exceptional child.

4. "Uncle Sprinkle" and his room at the Sprinkle Store were familiar to the writer in her very early childhood.

5. Medical men have pronounced this idea an absurdity until very lately when second thoughts are being given to conditioning of the male essence as regards conception.

6. The Driskill Ranch. Notes from obituaries.

7. A statement copied from W. L. Giles' accounts into Ledger No. 14, 1947-1952.

8. Cf. data, Ike T. Pryor, *The Trail Drivers of Texas,* p. 313.

RUBY ELIZA Jamison Swafford—the Cherokee.

ALFRED BEN ATKINSON, 1920

His whole way of life, in the eyes of Jane, who had married the well-to-do Virginia gentleman, James Corry Maxwell, put him out of the running for the prize of her baby sister's hand.[9] Frederic and Harriet would, as with their other children, offer their opinion, then leave Julie's choice to her own discretion.

Unexpectedly, just as the preceding winter began, one of Maria's grandsons, Jasper Newton Atkinson (John Pinkney's boy, called "Jap"), had come into the family circle on Walnut with his heart set on "going up the trail" like his Cousin Zackie when the spring drives started. Meanwhile, he would be content with odd jobs on the Rutherford plantation—like pouring excess from the churn down the streamlet the community called Buttermilk Branch, and taking the produce supply wagon (milk and butter) into town each day. His one adventure during the entire season was having the wagon slide off the road into the Branch on the morning Mrs. Erin Rutherford elected to ride in with him on a shopping expedition![10]

He escaped this humdrum routine in April by joining the Sauls and Armstrong outfit which was gathered at Round Rock (fifteen miles north of Austin on Brushy Creek), an outfit Zackie said "wasn't as rough as some of the others." Its destination was Oglellalah on the North Platte in Nebraska. Rough or not, when the nine thousand longhorns were delivered, young Atkinson was through with trail driving. Surfeited with thrills after at one time finding himself alone in a strange area with strange Indians and at another time riding for his life through a storm in front of stampeding cattle with the night illumined by lightning flashing from their horns, he sold his horse and saddle in Oglellalah, as did other cowboys, and rode home to Bellville on the train.[11] People on Walnut would not hear of him again until the eve of his son's marriage in 1926 to Frederic and Harriet's last grandchild, when a review of Alfred Ben Atkinson's genealogy disclosed their relationship.[12]

---

9. Ocie Cato, reminiscence to writer.

10. Told to writer by J. N. Atkinson in 1929.

11. *Dallas Morning News,* December 8, 1929, "A Ride for Life on the Oglellalah Trail."

12. Alfred Ben Atkinson was kidnapped in Shreveport, Louisiana, when he was eighteen months old, following his mother's death. His father left the baby with a seemingly kindly woman at a boarding house while he prepared to take the child to his wife's sister in Phoenix, Arizona. When he returned to pick up his son, woman and baby were gone. Despite search and advertising he heard nothing for more than twenty years. Meanwhile, the boy, named

It was in the following spring of '81, while more trail drives were getting under way, that horror struck. Harriet Emily came up from the Yegua country with her daughter, Belle, to visit the parental hearth. In the course of their visit, Zackie and John Payton, the latter in his fifteenth year, escorted Julie and Belle to the Odd Fellows' supper and ball in Austin, totally unaware of a measles epidemic raging in town. Texans, cut off from the mainstream of humanity by the isolation of

---

"Benjamin Schley" for the Spanish-American war hero but answering only to "Ben," grew up, soon satisfied he was no relation to the people claiming him. Threat of going to court to establish his identity forced from them his birthplace and his father's name of "Atkinson"—with the false statement that he had been abandoned on the streets of Shreveport, picked up and given to them under Louisiana's Sunday Adoption Law that no recording required (now abolished), which made the threat seem a waste of breath.

Coming to doubt the word of these semi-criminals, persons at the University of Texas, shortly before the "lost boy" was graduated an electrical engineer with Tau Beta Pi rating, instituted a search for the truth and found that older Shreveport residents well-remembered the baby's abduction and were able to put the inquirers in touch with Mr. Atkinson in El Paso. Just days before, a locally resident cousin from down Bellville way had called Mr. Atkinson and told him he was soon to hear from the boy lost at Shreveport! It was always said there were those in the family at Bellville who had Scottish "second sight."

Quickly, father and son were together in an Austin court room where, since the son had himself adopted the name "Alfred" and used it, with that of the kidnappers, on academic and Navy records, all of which must be changed, a new signature was established as "Alfred Ben Atkinson." Judge George Calhoun, in whose court this took place, when asked what should be done if the guilty parties were to try blackmail, saying "Ben" was not the Atkinson child, but another, the reply was "indict them for murder—the child they took has to be accounted for"—but he also warned that never again should there by any direct correspondence with them, or the courts would be powerless to act.

One other mystery was cleared up. Two days after World War I began, "Ben" left high school to enlist in the U. S. Navy. Two vessels were lent to England's Grand Fleet—the Lousiana and the Florida. A signalman on the Florida, he was denied entry to certain British treaty ports. British officials, asked why, said they didn't know what he was—but certainly he was no American! From his father it was learned that his mother, Lucy Lee Swafford, daughter of Isaac Scott and Eliza Ruth Jamison Swafford, from Neosho, Missouri, was, through her mother, of Cherokee descent. Slightly swarthy, her son had every mark of the classic tribesman: widow's peak with the hairline receding at each side, eyes a little tilted with dark smudges underneath called "eye-stains," dimples, cleft chin and the peculiar horizontal seam inside the

distance, had no immunity to the disease.[13]

John, the first to succumb, though quickly followed by Zackie and Julie, "broke out" one day in school. He went home, carrying the plague with him. Only two of his fellow pupils would escape the menace, and, in the family connection, only Amanda and her brother John, Jody's husband. Added to other worries was the belief that a similar contagion "settling" on Jim Maxwell in his Virginia boyhood was the cause of his continuing childlessness.

With Harriet and her daughter Mary, measles went into bronchitis. On Saturday, March 25, a wire reached Harriet Emily, back at home with Belle, and her neighboring uncle, William Gordon Bachman, that Harriet was dying. William Gordon, riding all night, arrived at daybreak to find Harriet gone. He brought word that Harriet Emily might herself be at the point of death. The same was true of Erasmus Sprinkle, and of Flora, with Sallie unable to leave her own bed to attend her. On the 29th, Mary Alford died.

Julie and Zackie, though both had been very ill, were the first to recover, as they had been the first to go down. Cremona Barr came and took her desperately ill father home.[14]

Of a sudden, the Alford children were scattered. Tom, the eldest, a grown-up twenty, went north with a cattle drive. Jennie, six years younger, moved over with their grandfather, whom the court appointed as their guardian, and her Aunt Julie. Her sister, Julia, sometimes called "Hallie," aged ten, went with the Maxwells, and Mollie, eight, to their Aunt Amanda. Baldy Samuel, sixteen, and Charlie, twelve, stayed at home, doing what they could to put in the year's crop. After harvest, Baldy would also go to live with the Maxwells until he, too, went "up the trail"; and Charlie with Dock and Sallie.[15]

Jim Maxwell, who had a soft spot in his heart for little girls, pretty

---

lower lip. Britishers, unsophisticated where American Indians were concerned, no doubt mistook him for a Eurasian—the Amerind having come originally from Asia. As very often happens when one is of Indian descent, no matter how far back, Ben's first child, Rose Ben-Marie, (who lived but three days) had the same distinctive features as her father.

13. For effect of smallpox and measles when introduced into new communities, see Bernal Diaz, *The True Conquest of Mexico,* or a recent work, *Plagues and People,* William H. McNeil.

14. This epidemic and the consequent deaths in the family are described in a letter in the Swofford file from Sallie Giles to her uncle, John Beneke Bachman, April 30, 1881.

15. Sallie's letter.

soon feeling that Eula Lee and Amanda's new baby, Dora Phoebe, were not much company for Mollie, who had been accustomed to older sisters, and that she needed to be with Alice and Laura rather than with "Amanda's big boys who teased her," got her transferred to Sallie's household.[16]

"Sleep," Shakespeare says in *Macbeth,* "knits up the raveled sleave of care." He could have said "time" in place of "sleep" and spoken even more truly, or he might have said "maturity."

Even so, a change came over the Jourdan Crossing on Walnut with the deaths of Harriet and Mary Catherine. A lacy iron fence enclosing a plot of ground in the pasture between their houses marked this change. There, instead of at Fiskville, they had been laid to rest, facing west to east, as was the Christian custom, ready to rise and face their Maker on the Judgement Day.

Each spring wild flowers would bloom inside the fence, remindful of a resurrection, but each autumn the hunter's moon, shining down through a patriarch cedar tree, would be remindful not of tenderness, but of Death, eternally searching for an unsuspecting prey.[17]

---

16. Statement of J. C. Maxwell to the writer *ca.* 1916.

17. The Jourdan family cemetery is on the Sprinkle Cut-Off Road that turns east from the Cameron-Dessau Road, just after crossing Walnut Creek.

THE MUSTANG

A spooked stampede:
Mad rush into the night—
Horns glinting white from the lightning,
Hooves out-thundering the thunder—
And a cowboy's bones
Saved from the crush of the flight
Where cow brutes jostle each other
In their senseless panic of fright.

Saved? Yes, by his mustang,
Roped on the mesquite range,
Brought strange from the prickly pear
To learn to cut out a yearling,
To head off a steer,
But full of a natural savvy
To turn, to twist, to rear
Away from the danger
Of cow brutes
Gone mindless with fear.

—M. Atkinson

*Mustang: Texanese from *mesteña* - Spanish range pasturage for undomesticated livestock; savvy: noun from the Spanish verb *saber* - to comprehend; Mesquite: Mexican Indian *mesquitl* - a thorn tree or brush growth common to Mexico and Texas.

The Austin American, January 29, 1958, says a law of 1852 required a passport for any cowboy going on the range to rope a mustang; cost of passport 50¢; tax on mustang taken 25¢; tax on wild mule taken $2. *The Austin Daily Statesman* is quoted as saying in 1881 ". . . there are thousands of fleet and game quarter horses in any part of the state."

PART TWO

The Cow Trails

It is said that civilization follows the plow; west of the Missouri the plow followed longhorn Texas cattle and Texas cowboys followed the longhorns.

—*''Brief'' by J. Frank Dobie, a typescript folded in the Giles copy of ''The Trail Drivers of Texas*

Red River Crossing was part of the age-old trail whereby vast shaggy herds came south to winter in Texas but, thanks to invention of the high power rifle that overshot Indian bows and arrows, by the time longhorns were started up the trail ''to meet the railroads,'' only bleached buffalo skulls marked the route of travel.

—*M. Atkinson*

JOHN AS "UNCLE BUD"

JODY

The Raveled Sleave*

That's all chimney corner law, you just as well put it in your pipe and smoke it.

—Frederic

On October 20, 1881, Julia Elizabeth Jourdan and the cowboy, Osceola Claiborn Cato, were married. Walnut Creekers observed the Scottish custom of celebrating with a wedding feast at the bride's home, followed next day by an infare dinner at the groom's house.[1] It was on the latter occasion that (presumably) the formal wedding portrait was made, the bride in her "second day's dress" often posed standing, a hand on her seated husband's shoulder. It was at this time that Jane Maxwell, catching Ocie in the hall and alone for a moment, tartly inquired, "Now, sir, you have got a wife, what are you going to do with her?" to which he replied, after a moment's consideration, "Support her, I hope."[2]

With a shrug and a sniff a little too lady-like for a snort, his inquisitor marched on to more congenial company.

Eight days later announcement came from Elgin of the birth of Georgia Elizabeth Fisher, marking the advent of a new generation. The baby was the daughter of Frances Priscilla and James Vaughn Fisher and the granddaughter of the lost Confederate soldier, George W. Jourdan.[3]

In this decade, Fiskville, with Zimmerman's Store its focal point, was the principal Walnut Creek gathering place. The store's first floor

1. Infare: according to Webster, an "entering in"—groom's reception for the bride.

2. Told to writer by O.C. Cato.

3. Data from Georgia's daughter, Mrs. Isabelle Dubar, Elgin, Texas.

*Sleave: a skein of raw silk.

housed the post office and a saloon. The upper floor accommodated social affairs. Dr. A. S. Shannon lived in hailing distance; the blacksmith shop, proprietor W. E. Younger, and the sorghum mill (for molasses), proprietor John H. Mulkey, were handy when picking up the mail that came out from Austin by stagecoach three times a week. So also was "Grandmother Dean's" carpet loom, on which she made rag carpets and rugs for other ladies. "Wash," George Washington Cazine who had been the fiddler and belonged to "Aunt Betsy" Cazine, one of Sallie Giles' best friends, now owned and operated the gin and was among the district's largest land owners.[4] The *1874 Texas State Gazette* lists his land spread (which Eugene's Memoirs say extended to the Sinnet Musset place) as 725 acres. The *Gazette* says the stage stand was named in 1872 for New York born Josiah Fisk.

Fiskville was scattered on the upper reach of Little Walnut, a westerly branch of Big Walnut, or Walnut Creek proper, and had the advantage of being well out of Black Land mud when it rained. Its white soil, called "Austin chalk," grew pygmy cotton, two bolls to a stalk. This Black Landers ridiculed as "bumble-bee cotton," saying it "grew only knee-high to a bumblebee."[5]

When traveling evangels became aware of advancing frontier prosperity in the wake of Comanche withdrawal from the hills after reactivation of the federal forts, Fiskville, offering camp-site advantages for "protracted meetings," had a head start toward first place among the county hamlets clustered in a semifeudal domain around Austin. Which was to reckon without taking into account ambition stirring in a very young man over on Big Walnut, William Braxton Barr, Erasmus Sprinkle's second grandson.

In October of 1882 Julie and Ocie's first child, Ethel Chase, was born; and in December Frederic paid off four promissory notes totalling $2,656.50, with 10% interest, to E. Reeves for 253 acres.[6] It

---

4. Cazine: Ledger No. 9, 1937. Deed Book 67, pps. 295-298; Deed Book 79, pps. 639-649. The *Texas State Gazette,* 1874, credits G. W. Coryine (*sic.*) with 725 acres which Eugene says reached to the Musset place at the point where an old Austin road crossed Braker's Lane. According to Eliza Norwood Shanks, Alice, his only child, "ran away" and married Eliza's cousin, Alf Raney, who is listed in the *Austin Directory* (for Sprinkle) 1898-99. She describes Cazine as being quite black but his wife, Minnie, as "light" complexioned.

5. Explanation from writer's father.

6. Deed Book 47, p. 580.

was his second and largest buy since the close of the war, the first having been 114 acres acquired that same year at a tax sale.[7] Meanwhile, he had given fifty acres each to Sallie, John and Zackie.[8] In March of '83, he deeded a hundred acres to Jane.[9] He had been selling all along to German neighbors, northward, and to Black folk settling on the east side of The Field. Beyond the Blacks, he sold to A. J. Almquest, prosperous among incoming Swedish settlers.[10] But over and above his land deals, he was beginning to think in terms of Austin's expansion and of broader interests for himself than real estate and a plantation commissary.

Death did not wait too long to make another claim in the absence of medical understanding, common to the period. John and Jody's eight-year-old Eddie died April 13, 1883, after a month of agony with what would later be assumed to have been a burst appendix; it was then called, for want of better terminology, "inflammation of the bowels." Five weeks later their third son, William Pannell, or "Billie," was born. Like John, Jr. with Sallie's Eugene, Billie had a twin cousin—Jody's sister Fannie's boy, William, called "Buddie" by her other five children, who were girls. Fannie, married to a tenant farmer, Flynn Garrison Thurman, had made her home in The Field, to Jody's comfort of mind.

Late in the year, Frederic put down $2,000 cash and gave notes for $4,000 at 10% interest to be paid within twenty-four months on Austin property at the corner of Congress Avenue and Live Oak (now 2nd Street).[11] His legal advisors were the Sneed connection; not the judge (who died in '79), but his son, Thomas, and his son-in-law, Calvin Goodloe. They knew Frederic well and knew he did not always abide by their advice.

> "One time," Goodloe would relate years later, "the old man came into our office to find out what he could do to recover a thousand dollars he had with the Palm Company on Congress. The company was in bankruptcy.
>
> "We had to tell him he had no legal recourse under the circumstances. He didn't say anything. He left, but we knew that wasn't the end of it. There was talk that those Palms (not Sir Swante Palm) made a practice of bankruptcy. Two of us young fellows followed him to see what he would do.

7. Deed Book 47, p. 385.
8. Deed Book 50, p. 31; Deed Book 49; p. 634; Deed Book 50, p. 90.
9. Deed Book 55, p. 249.
10. Deed Book 55, p. 278.
11. Deed Book 58, p. 342.

> "He didn't go down the Avenue. He had already been down there and got the door shut in his face. This time he turned into the alley. Just before he knocked on the back door he pulled out a pistol that looked two feet long to us young fellows watching from round a corner. When the door opened a crack, he stuck the muzzle in, then he stuck his foot in, then he went in. It wasn't more than a minute before he came back out, a roll of money in one hand, the pistol in the other. He stuck the pistol in his pocket and went off, flipping the bills, counting them.
>
> "We went back to the office and we never heard any more about his account with the Palms."[12]

As time passed, they were to hear quite a bit about a business he was proposing to start on the northwest corner of Congress and 2nd Street.

Erasmus Frederick Sprinkle was also changing business interests. His enterprising young grandson had acquired (at a tax sale, it is said) a spread of land around the commissary. "Brack," aged twenty, was just married to Matilda Birdwell, fifteen-year-old sister of his brother Hayward's wife, Minnie. The Birdwell girls were from Buda, on the San Antonio side of Austin. Mr. Sprinkle ("Ras" to his immediate friends, "Uncle Raz" to his nieces and nephews at Hornsby's Bend) turned the commissary over to Brack, who got himself appointed postmaster for the district and the district designated "Sprinkle" in 1884.[13]

In Austin, a cornerstone for the structure Frederic was building would be marked above the date

F. JOURDAN
F. SPRINKLE

Perhaps it seemed euphonious to use two "Fs" even though "Frederick" was only Erasmus' middle name. No evidence has appeared to show his direct concern with Frederic's project, yet one is compelled to believe, from the use of the two names, that a partnership was at least contemplated. (The writer, when a child, frequently visited this corner, intrigued that, like the cenotaph in the cemetery, it memorialized a grandfather she could know only through hearsay; for her, the double lettering signified a harmony that must have existed between him and the old gentleman with a snowy beard and intensely blue eyes she had known when very small and had been taught to call "Uncle Sprinkle." When she looked for it in 1965 the two-story building at the northeast corner of Congress and 2nd Street was gone—demolished in favor of a parking lot.)

---

12. Recounted to Jody in the writer's presence, Houston, Texas, *ca.* 1910.
13. Ledger No. 9, 1937.

Trouble, unconnected with the new venture, lay in wait for Frederic. Austin had lawyers other than the Sneeds. One was Taylor Moore, Lucy Payton Jourdan Moore's stepson. Frederic knew perfectly well Texas' community property law, which vested one-half of The Field in Harriet's sucession, but by tacit consent the land had remained undivided except as he chose to distribute acreages in accord with where his married children had their homes. It was to his regret that Harriet Emily was gone with the disgruntled Waldemore. Lucy's sons, James W. and George W., brought up at Bastrop, were now, like Brack, grown young men, and like him, uninterested in cotton planting. Whether anyone other than themselves and Taylor Moore knew that rising young lawyer was holding their note for $300 borrowed at 10% interest is a matter for surmise, but when Moore, with a power of attorney, presented his step-brothers' claim to their father's part of their grandmother's estate,[14] Frederic knew the jig was up so far as avoiding a partition was concerned.[15] He told his affronted children that what he or they thought was right or wise to do about The Field would cut no ice in court: Harriet hadn't made a will and their argument as to what Mother would have wanted Pappy to have or to do "was only chimney corner law—they just as well put it in their pipes and smoke it!" He had no intention of himself "going to law" when he was certain to lose, so, if the legal machinery were set in motion, let it proceed without friction according to the Texas rule of descent and distribution.

On the 1st of May, 1885, Travis County's probate court determined each of the nine heirships for Harriet's children to be 63 acres of The Field and a ninth of the cedared block across Walnut which was slightly more than 100 wild acres good only for fence rails and firewood, this block to be partitioned by common consent. Soon as this determination was final, Frederic mortgaged his remaining 541 acres in The Field to Brown, the English land broker in Austin, for $3,500 at 10% interest, the principal due May 1, 1888, and he validated this mortgage by moving into quarters above his building; the loan is part of the instrument recording the partition.[16] The next city directory records him as a dealer in flour, but there are indications that he was stocking lines for a department store.[17] On July 12th, ten

---

14. Power of attorney to E. T. Moore, Deed Book 63, p. 559.

15. Frederic's reaction recounted by Jody.

16. Deed Book 64, pps. 224-225. See Appendix D for division of Harriet's estate.

17. A receipted payment of $1,039.25 to Moses & Son, Clothiers and

weeks after the partition, Harriet Emily Alford-Waldemore died. Laura set down the date but no record was made of where her aunt died or was buried.

The year 1885 was a year of many conclusions. Zackie married Isabelle Daugherty—the last wedding of his generation. Amanda's sixth son, Bradford Norman (named for her and Sallie) was born to complete her family. Ocie Cato made his final drive north. Jim and Jane Maxwell moved into Austin with Jim going into Wayland's wholesale grain and produce business.

The next year Zackie and Belle's son (to be their only child) was named Eugene Edward. Jody said the baby was to have been called after her lost Edwin but at the moment it was not understood the spelling should have been "Edwin." It would make little difference for this his thirty-third grandchild (fifth from the end of that page of Biblical inscriptions) was never to be called by it, anyway. A pretty and an active youngster, he so captivated Jane's fancy she wished her father to make him sole heir of the remaining estate, thus establishing him as a landed gentleman after the European fashion. Jody thought Jane's plainness of feature, together with her childlessness, made her peculiarly susceptible to such infantile attractiveness, especially in a boy to carry on the family name.

Frederic had no such notion. Charmed as anyone else with a child spinning about so agilely, he declared him to be "a regular little jeter" (or so the family understood him to say) and fixed on him a title which first was "Jeter" and then "Gete."[18] Apparently "Jeter" was from *jeté*, a French term in ballet dancing. Frederic was aging. There may have been moments when his mind harked back to his own youth and his French ancestry.[19] Or he may never have intended living in

---

Gent's Furnishers, 613 Congress Avenue, dated January 28, 1888, to discharge a note given by "F. Jourdan and others" on February 14, 1885, appears in the Giles papers. This payment by Z. P. Jourdan on behalf of Frederic's estate, plus the writer's memory of a fragment among her father's papers concerning a deal in "laces and fine fabrics," leads to the conclusion that much more than household staples were intended to be dealt in.

18. Reminiscence, Laura Giles to writer.

19. Frederic was a man who did not waste words on his past. It has not been possible, so far, to trace his lineage. Beyond the two facts: that he spelled his name always the French way and gave his birthplace invariably as North Carolina, there is only a slight clue in his having said he named Amanda "for a little girl he used to know." This is all there is to go on except a story told to the writer in 1935 by William Forrest Jourdan (born in Tishomingo County, Mississippi) who was then county superintendent of schools at Del Rio, Texas.

town the year round, only declaring the building on Congress and Walnut as his residence so as to release the Walnut Creek homestead for security in the loan to finance the Austin venture when the forced division of the plantation cut his resources for expansion in half.[20] Whatever he had in mind, his plan seemed to change, for in June he rented the building for $50 a month to J. P. Sprinkle (who was in the insurance business, commuting from Hornsby's Bend) and went back to Walnut.[21]

He had lived a full life of three-quarters of a century when he died in the Indian summer of 1887. No more ready for death than

---

He said that, arriving in Texas after 1901 and learning of a Jourdan family at Austin, he visited Z. P. (then county treasurer) at Austin.

After this visit he wrote his "youngest aunt," asking if she "ever heard of a Frederic Jourdan?" Her reply, as he recited it:

> Her brother left home and was not heard of for several years (not an unusual occurrence with young men of that time when things did not go to suit them at home and communication was poor). Finally a letter came saying he was married and living in Texas. Four members of the family wrote an immediate answer, shedding tears on the paper, begging him to come home.
>
> It was a year before they heard again. Then another letter arrived saying he had got hold of some land and some Black people and he thought he would stay in Texas. This time all the family signed the answer, pleading, "Dear brother, please come home—the Indians will kill you down there!" and blotted their signatures with their tears.

"But we never heard any more," the aunt's letter concluded and we just supposed the Indians did kill him."

"And I wrote my aunt," William Forrest said, "there are a damn sight more Jourdans than there are Indians in Texas now!"

He added that the original Jourdan of this line came over with Lafayette, went back to France and sent his two married brothers over to take up land. The head of the family, as he gave the story, then unmarried, was a man of title, one Jean Baptiste Jourdan. But inasmuch as *Les Combatants Francais de la Guerre Americaine,* 1778-1783 (the roster of the expeditionary force) gives forty-one Jourdans—nine of them named Jean, three named Jean Baptiste—and since William Forrest died the next year (1937) before the writer had enough research experience to ask further questions, this story has to stand as he told it and as the writer remembers hearing it.

20. See fn. 14, Chapter Three: "Below the Balcones" for the Homestead Law.

21. Deed Book 64, p. 405.

Harriet, he had made no will. Five weeks before succumbing to a painful congestion in his head, he receipted for William Boyce (youngest son of Elizabeth Boyce) payment for 50 acres of land over toward Gilliland,[22] bought in 1883.[23]

Harriet Emily, George W. and Mary Catherine had been his children born east of the Mississippi. Their going and his, following Harriet's, broke the family's blood bond with the Old South. Straightening out any tangle left behind became the responsibility of the Texas-born once they had erected a modest cenotaph inside the lacy iron fence under the patriarch cedar tree.

---

22. Deed Book 75, p. 309.
23. Deed Book 55, p. 387.

9

Miss Sallie's Boy

Oh, for boyhood's time of June,
Crowding years in one brief moon
When all the things I heard or saw,
Me, their master, waited for.
—John Greenleaf Whittier, "The Barefoot Boy"

Sallie's boy, Eugene, was not yet out of the kilts he and his cousin John and other little boys wore till they were four years old when he drew from his grandfather's tall hat slips of paper marked one to nine designating for his mother and his aunts and uncles the 63 acres each was to inherit, by court order, from his grandmother's estate.[1] All the swapping about they would do thereafter concerned him not a whit. The baby in his family, the beloved boy, his was the freedom to explore his universe and rarely was an amusement of his decreed out of bounds, though it could happen.

One such occasion he would ever remember caused his mother to forget her own upbringing by Harriet who did not believe in physical chastisement as a means of setting a child's feet on the path he or she should tread—not even when Jane stubbornly replaced her cornshuck dolls and cockleburr furniture on a mantelshelf Miss Mary and Miss Em'ly had just cleared in anticipation of Sunday afternoon callers.[2] Eugene was past his third birthday when someone gave him a toy hatchet.

In Sallie's henhouse was a pen of baby turkeys needing to be frequently fed curded clabber sprinkled with black pepper. Turkey chicks brought off the nest in domesticity must be coddled—kept dry and warm on a special diet through their first two or three weeks lest

1. E.V.G.'s reminiscence to writer, expressed with a humorous hope that he would never be blamed for any results from the drawing.

2. Jane's reminiscence to the writer.

they wilt and die like uprooted flowers. During the school term Alice, Laura and Mollie left home early and stayed late at the Gilliland Academy. Sallie, with Flora, Maggie and Eugene to care for, and the little turkeys as well, had her hands more than full. Toward noon one day (as it happens to every mother at times) a feeling came over her that the world was unusually quiet; the daughters beside her were for once soundless, absorbed in basting their bright bits of cloth together with half-inch stitches. Without their noticing, she slipped out the back way to see what the boy was doing. It may have been primal intuition that led her to the henhouse on quiet feet; at any rate, what she saw there did not chill her blood—it made it boil.

Under the henhouse shed, young Eugene, toy hatchet raised, was stealthily creeping on another little turkey while round him in the straw lay, beheaded, half the brood from two settings of eggs! Had one of his sisters been home from school to make the gruesome discovery, break the news and intercede for him, Sallie's wrath might have cooled before being fully vented; as it was, the sanctity of her henhouse was forever impressed on her son's sensitive rear.[3]

Another experience of great trauma (though it was entirely emotional) occurred on a day when he went with his father to the post office in Zimmerman's store. He had but recently graduated from kilties to knee britches and brasstoed boots, of which he was inordinately proud. A Black woman he did not recognize, appearing out of nowhere, gathered him to her bosom and covered his astonished and reluctant face with kisses while crying to an interested and onlooking world, "If this here ain't Miss Sallie's little boy!" Half a century later he would write that it took him days to recover his manly self-esteem.

A third time he bewailed his fate was after reaching an age of responsibility such that he was allowed to ride bareback through The Field to his Uncle Jim Maxwell's gristmill with a sack of corn only a little smaller than himself balanced behind him. Either the corn shifted or he did—or the gentle old nag he was riding shied a bit—he was never sure which—at any rate, he and the sack slid together into the soft dirt of a turnrow, out of sight or call of a house and out of reach of mounting block or tree stump. He was fortunate in that before his predicament became unsupportable his Uncle Bud came riding along on an errand of his own to lift boy and grain sack back into place, the while punctuating his kindness with some carefully

---

3. All incidents relating to his childhood are copied almost verbatim from the Memoirs.

chosen comments on his nephew's need to improve his horsemanship.

It was not too long after this that the same Uncle Bud rode up to the Giles' front gate accompanying twin-cousin John on a pony caparisoned with a brand new saddle, bridle and bright blue saddle blanket flaunting a white star. Such a display on the part of a peer was too much for a small boy recently twitted by the peer's father for being dumped in the dirt from the bare back of an ornery old mare! Without taking a second look, Eugene dashed into the house shouting for his mother, literally screaming to her that if John had a pony and saddle he must have one, too! Presently pacified by promise of a matching outfit when his parents should be in Austin again, he recovered from his tantrum sufficiently to come back out and take an interest in his cousin's new trappings.

Aside from experiences like these, his boyhood, according to his Memoirs, was one of untrammeled joy. In the great outdoors, butterfly and moth cocoons festooned the undersides of fence rails while corners where rails overlapped housed bird nests with eggs or young to be counted and reported at the supper table—a triple exercise in nature study, arithmetic and the art of conversation. The art of self-defense also entered into this casual accumulation of knowledge for he became well acquainted with the ways of wasps and bumblebees.

The bumblebees fascinated him with their artisans' work, boring perfectly round tunnels the size of his finger, some almost from end to end of a rail. Experience taught him that none would attack if merely observed, that white-headed ones could be safely caught and whirled on a string, but black-headed ones were vicious if disturbed: like the wasps that attached grey clusterings of paper cells—some clusters as large as a grown-up's hand—to rails and posts by tough half-inch long paper stems. Wasps were usually the small black-striped yellow jackets though sometimes present was the larger red hornet type. All these fence residents were summer phenomena.

When autumn came he scanned the sky for bird flights coming down from the north and in spring for flights returning from the south. Always he searched the grasses for the year's first flowering starlet. Thus, like an Indian, he learned the seasons.

His most exciting moments, outside of circus day, came during the grain harvest, during hog-killing time and on Sundays when neighbor "Cam" Hill from Gregg's Corner preached damnation to sinners at the Burditt Schoolhouse, a little south of the Jourdan Crossing on Walnut. "Brother" Hill was a layman of the "Christian" persuasion—a denomination founded by Alexander Campbell, whose followers (to their great annoyance) were called "Campbellites" by

Methodists and Baptists who resented this preemption of "Christian." Brother Hill's hellfire exhortations were delivered without other compensation than a release of his own pent-up emotions and an awareness of the lift given the imaginations of his listeners, including a search of their consciences.[4] The school house, built by public subscription, had the month's four Sundays allotted to four separate sects, with the rare fifth Sunday allotted to a congregation numbering a single family. Everyone was, of course, welcome to attend any and all services.[5]

Custom was for men to sit on the left side of the building, women on the right. Eugene, proud of his first breeches, was innocently following his mother when he felt himself grabbed by the seat of his pants and looked back to find a grey-haired lady vigorously nodding him to a place across the aisle. It increased his sense of importance after he thought about it, and he never made that mistake again; but very often, when Brother Hill was at his best, his boy's soul seemed to shrivel within him—for had not his many yieldings to temptation already put salvation beyond his grasp?

Wasn't there that afternoon in the henhouse when he had beheaded the baby turkeys? And that morning in the orchard when he nibbled the peaches he had been forbidden to pluck, leaving their fleshless seeds suspended from his father's prized experimental tree? "Be sure your sins will find you out!" Brother Hill thundered. Well, they had, and there might be others, forgotten or unrecognized at the moment of commission, eternally waiting to pounce in the hereafter since the stentorian voice echoing from the rafters assured him that "up there" his every act was being recorded to confront him, "come Judgement Day!"

Thus awful was his concept of his final end and hopeless any idea of avoiding it, yet it was seemingly so far off that, once out of his Sunday best and into his play clothes, he forgot all about it in a game of tag with his colored friends or a chase after Ole Lil—a genius at lifting fence rails to escape from the cowpen.

If anything could outdo an evangelistic preacher at making itself heard, it was an engine powering a threshing machine. Eugene,

---

4. Alexander Campbell, a Virginian, derived the Christian Church from Presbyterianism prior to the Civil War. His follower, Alexander Campbell Hill (popularly referred to as "Cam") was a family connection through marriage, inasmuch as his wife, *neé* Lou Maxwell, was a first cousin of Jane's husband. Cf. fn. 12, Chapter Twelve, "Sprinkle."

5. Unfortunately, the Memoirs do not name the sects.

granted the privilege of pulling the whistle cord, would hang on the yard gate each harvest season, watching approach of the threshing outfit on its way to The Field, savoring the moments when his ego would be as fulfilled as Brother Hill's in delivering a hell's fire and brimstone sermon.

First came four or six yoke of oxen, pulling the engine; then four more oxen hauling the threshing machine; then another yoke drawing the water wagon, and, finally a crew of eighteen or twenty Black men on their ponies. The outfit belonged to George Washington Cazine (colored) from over Fiskville way. Both men and animals would expect food at the noon hour, which meant Aunt Cynthia Hill (also colored) had been invited to help Sallie and the girls in the kitchen while Aunt Cynthia's boys fetched and carried, especially water, for the sweating men, their ponies and the oxen. "Wash" would take a tenth of the threshed grain in payment for the operation.[6]

After wheat and oats were sold or stored, corn gathered into cribs, pumpkins and cashaws from between the corn rows piled for winter eating and the last cotton scrapped from the dead stalks, circus day arrived. For Eugene, this meant being awakened before daylight to start the two-hour drive to town in a wagon bed padded with quilts since the cotton wagon was the only vehicle of a size to accommodate the entire family together with a day's provender for both them and the horses. So early a start meant to secure a place on Congress Avenue where no part of the grand spectacle would be missed—from the ringmaster in his top hat to the steam organ (the Calliope—always pronounced cál-i-ōpe, to rhyme with hope) at the tail end of the parade. It also meant getting back after dark, surfeited with wonders, peanuts, popcorn and cotton candy, and being carried into the house sound asleep to wake in the morning, still with one's clothes on.

Only Christmas Day surpassed circus day in anticipatory excitement. Eugene's two complaints about Christmas were that, though he sometimes heard Santa scrambling back up the chimney, he never could catch him coming down, and that his stocking was shorter than any of the girls'—none of which he was allowed to borrow because Santa would not understand the exchange. By the time his father finished a Christmas Eve reading of Scrooge's conversion from monster to benefactor of another little boy named Tim, Eugene was drowsy enough for bed, where Flora and Maggie were already gone. This left Alice and Laura and their mother and Mollie to hear other stories the head of the House of Giles would read beside a candle-lighted cedar

---

6. See note 3, Chapter Eight, "The Raveled Sleave."

tree in the corner. The sooner you are asleep, they told the boy, the sooner you will wake up in the morning to see what Santa has left under the tree for you—an equation he never quite understood.

Hog-killing meant another arrival of Aunt Cynthia and family, as well as of the ancient bachelor, "Uncle A" (Aaron White), who had been one of Frederic's Blacks and who was now living alone in a log cabin up Braker Lane where the colored Methodist church, Mount Salem, faced, cater-cornered, the colored Baptist church, Mount Zion, at the crossroad of the lane and the road leading from the Musset place past Fiskville to Austin.[7] Patriarchal over the festival was Dock in his long overcoat cut like a frock coat, his lengthy brownish, always slightly curly beard and his "Darby," conspicuous throughout the countryside among rolled Stetsons like Ocie's and Zackie's (the mark of the cow man) and broad-brimmed black planters' hats like John Sr.'s and Jim Maxwell's. He had another distinguishing feature, far more unusual but unnoticeable except at close hand: he had one brown eye but the other was equally divided crosswise—half brown, half blue. The writer well remembers from her infancy his coming calling on the family of an evening and her sitting in his lap, pondering over the strange eye, with little golden-brown flecks in the blue part, till she fell asleep.

A shoat (half-grown pig) would have been killed on the season's first norther to provide fresh meat for a while, but not until the first real freeze-up would hilarity reign around outdoor fires where people, bundled against the cold in two or three coats and perhaps as many hats, one above the other, rallied one another on their unaccustomed attire as they waited for the slaughter of a dozen or so porkers to start the great ritual of preparing a year's supply of ham, shoulders, side meat (bacon), sausage, souse (also called "hog's head cheese"), liver pudding and lard. Souse and liver pudding were traditional, technique of the former having been handed down by Harriet and of the latter by Nancy Caroline, Sallie's mother-in-law.[8]

---

7. Names and denominations of these churches were supplied by Mrs. Mary Eliza Norwood Shanks. The writer recalls their juxtaposition with Uncle A's cabin. Dock Giles' notebook for 1892 has references to Aaron White, deceased: (1) Debts owed by the estate of the deceased: Dr. Bradfield, \$58; Dr. T. D. Wooten and Sons, \$2; Dr. A. S. Shannon, \$4.45; Monroe Miller (mortician) \$30. (2) Property belonging to estate of deceased: January 16, cow and a calf.

8. Liver pudding required a heart and a liver to be boiled, then the heart diced and the liver crumbled. A pound of leaf lard was also diced and these ingredients mixed with corn meal, the former equalling two-thirds of the bulk.

The ritual, though its thrills were exhausted in the first two or three days, extended over weeks, with the fresh meat lying in salt for a month and a half, hence the need for freezing weather, after which a smudge fire was built in the smokehouse with a care approaching reverence—it must be of certain woods for flavor, it must never be allowed to go out, nor must it ever blaze—over which the various cuts of meat would hang till "cured"; then they were packed in wood ashes (to keep out "skippers"), or in cottonseed, whence they were brought forth when needed for the dining table.[9] Even the making of lye soap to wash clothes was a by-product of hog-killing because of the fats combined with balls of lye in what was also a family recipe.[10]

Much as Eugene enjoyed fresh pork and the sausages for which Aunt Cynthia prepared the casings (he considered cased sausage far superior to the sausage his Uncle Bud's folks put up in narrow muslin sacks to slice off in patties); for him the highlight of the whole proceeding was his vested claim on the hog bladders. These were blown up while green through a quill or a bit of cane inserted at the opening, tied off, and the bladder covered with the top of a stocking. Mounted on a stick such a bladder became a court jester's sceptre, or, ballooned on a string, it could be bobbled about much like the harmless white-headed bumblebees he and his dusky playmates sometimes harnessed for the fun of it. Balloons thus achieved could be enjoyed for days, till they shriveled. All his lifetime he would hold

---

This mixture was spooned into squares of muslin and the squares tied in bags, being sure to leave double room for the meal's swelling, after which the bags were dropped into boiling water kept boiling until the meal was thoroughly done.

As needed, the muslin was peeled off and the "pudding" sliced and served hot as it came out of the pot or sliced and eaten cold or fried.

9. Hams smoked in this manner were spoken of as "grey" hams and the technique as "Indian." Virginia hams, or "pink" hams, were cured differently, it was said, by the use of saltpetre. Uncle Jim Maxwell's table was set with pink hams.

Doris Riedell, who operates the Jourdan-Bachman Pioneer Farm, spent much time in childhood with her maternal grandparents at Bee Caves. Her grandmother, Roxie Tombo (Mrs. John Hickson) was Tonkawa —and the smokehouse fires were never stoked with other than corncobs, except that a stick of hickory or some special wood might be added for flavor.

These hams yielded the thin, speckled gravy, so delicious on hot biscuits, that Ocie Cato saw as "brindle gravy."

10. Soap, according to the writer's memory, was made in a black iron washpot over an outdoor fire.

DOCK GILES with cow—Dock in hog-killing garb.

them superior to any "boughten" variety of balloon, just as he would hold chewing gum made of stretcher-berries and gum-elastic gathered along the creek superior to modern bubble gum.

Each return of spring started the cycle of his pleasures over again: green grape pies, dewberry pies and swimming in the "round rock hole" down by the crossing with Dick Bradshaw and Ned Jesse—Black boys who "lived on the place."[11] Even before spring of 1888 came in one finds him searching the grasses for early bloom—a diary note Laura made on February 16, together with a star of tiny faded petals, says, "This little flower was presented to me by Eugene on his eighth birthday."[12] In the summer of that year he would be considered old enough for formal schooling and once an educational routine was begun, his unbounded freedom when he felt himself monarch of all he surveyed would have come to a natural end.

---

11. A note in the Memoirs says the "round rock hole" was long since filled by the wash of debris. At no time could Eugene have said, like his cousin John Lafayette Alford, who had been a boy twenty years earlier, that "the creek had a rock bottom," or "the water was so clear you could see all the fish. We used to go seining and catch them by the towsack full." Letter, November 11, 1939.

12. This habit continued through adulthood. One diary notes with surprise a single bluebonnet found in January. Ordinarily March 15 to May 15 is "bluebonnet time."

JOHN IN his kiltie

EUGENE IN his kiltie

10

The Walnut Creek Avunculate *

> *"Avunculus" (from the Latin - "uncle").*
> *The special relationship among some tribal peoples between a nephew and his maternal uncle; the authority of a man over his sister's children and the reciprocal rights and prerogatives associated therewith.*
>
> —*Webster's New International Dictionary*

Since Frederic died intestate, as had Harriet, his heirs, having cut their teeth on court procedure in her case, gave to William Leroy Giles, John Bachman Jourdan and Zachariah Pannell Jourdan a power of attorney enabling them to act in concert as executors of his estate.[1] "Avunculate" is a good enough term for the authority, in keeping with the state's law of distribution and descent, which was vested in these three through this action by seventeen interested parties: four sisters, two brothers, five nieces and six nephews. At the same time, James Corry Maxwell replaced Frederic as guardian of the Alford minors: Charles, Julia and Mollie.

Once again there was a drawing of paper slips numbered one to nine and again there was trading to rearrange boundary lines. In the new division, the "home place," with the spring at the head of the little box canyon above the creek crossing, fell to Sallie. It would thereafter be known as the Old Place. She would never live there, preferring her own house and garden a mile or so up the road to the northeast. Within six weeks John sold the adjoining strip eastward to Jane for $1,125 cash and she resold it to Zackie for $75 more, to be

1. Deed Book 79, p. 209.
2. Deed Book 80, p. 397; Deed Book 55, p. 249.

*Avunculate: a matriarchal society with uncles rather than fathers heading the household.

paid at his convenience with 10% interest.[2] She and her husband had moved into town about the time Frederic did, taking Julia Alford with them. James Maxwell, the Virginian, a merchant at heart, though he listed himself in the 1887-1888 directory as a farmer, had acquired an interest in the John A. Wayland wholesale and retail grocery business, where Charles Alford, beginning as a clerk, soon was launched on a career that would distinguish him eventually as Austin's foremost grocer.[3]

Before the agreed-on division of the land could be affirmed, the administrators, according to an invoice of Frederic's estate, faced an indebtedness of $8,319.43, one-third secured in part by the lien on his remaining acreage for the money which had enabled his start in business at 2nd and Congress.[4] Family consensus was that with cotton coming to the fore, it were best to hold on to the rich black soil at all costs.[5] Refinancing in January of 1888, making Frederic's obligations into a single debt secured by the Austin property, allowed The Field to be tilled for the coming cotton season.[6]

It was with relief that, in 1890, the three administrators finally affixed their signatures to a quit claim on the Jourdan-Sprinkle building in return for surrendering to Franz Fizet and C. H. Miller, Brokers, the notes against it and receipting for those gentlemen $10,000—$2.000 of it "cash in hand paid"—a remainder to be divided, like the land, in nine parts, then three of those parts redivided amongst the two sets of Alford nieces and nephews and George's heirs.[7]

James Jourdan, as soon as he legally could, which was in the spring of '88, arranged for his own and his sister's and brother's holdings from Frederic to be converted to cash, just as he had done with their holdings from Harriet which they sold to their Uncle Zack.[8] This time, with the exception of a tiny strip of 4¾ acres, they sold to an outsider, one Parrish. Squarely in the middle of The Field, this 40 acres would soon be retrieved.[9]

When the hunter's moon rose in October of '88, James, just under thirty, sat coughing himself to death of pneumonia in his Aunt

3. *Austin City Directory,* 1887-1888.

4. Deed Book 64, p. 227.

5. According to Jody.

6. Deed Book 81, p. 517.

7. Deed Book 98, p. 56. See Appendix E.

8. Deed Book 63, p. 560; Deed Book 81, p. 528; Deed Book 80, p. 412.

9. This deed was not found by writer, but the land has long been family owned.

Amanda's house on the Mountain Road. Half delirious, he kept murmuring, "I'm a gone-up chicken! I'm a gone-up chicken!" till he sank into a coma and died. The scene would stay forever in his aunt's memory.[10]

Laura Giles recorded the date of James' death in her diary. The following line reads, "George Jourdan died ____." Perhaps she was uncertain and put the book aside intending to fill in the date of George's death later, then forgot it. At any rate, neither brother signed the quit claim to Fizet and Miller with their sister Fannie Fisher, and her name only is on the reckoning sheet in the Giles Papers showing the division of the $2,000 remainder.[11] Nor does anyone know where either brother is buried, though it has been assumed that James is buried in the Fiskville cemetery.

Like James, John Lafayette Alford and his sister, Baldy Belle, sold their inheritance quickly to other family members.[12] Both were married into German families "down the country," Jack to Augusta Heine, Belle to Oscar Swartz. They would continue to live where they were; that is, after they took a flyer, in company with the despised Waldemore, at some highly touted-up land in Coryell County where a drouth promptly wiped out their speculative down payment. "Man," they would console themselves for their loss, "cannot control the elements."

By the opening of the 1890s, The Field was settled down to an expanse of cotton farms extending from what is now the Cameron-Dessau Road on the west to the main leg of the Cameron Road on the east—a labyrinth of turnrows bewildering to the uninitiate who might be bold enough to leave either of the beaten paths in search of a short cut to the other.[13]

---

10. Recounted to writer by Amanda.

11. Deed Book 98, p. 56. Appendix E.

12. Deed Book 80, p. 489; Deed Book 80, p. 397. Interview with Mrs. Emily Swartz Rutland, Kingsville, Texas, 1977. See Appendix F for division map.

13. The "Cameron Road" has itself become a labyrinth—due to its many changes—and can easily lead the unwary right back to his starting place!

11

Rose Cottage, Nowhere

Three things never stay to wait man's pleasure: fading flowers, rivers that seaward take their way and passing hours.
—Japanese Poem

Sallie's house faced west, looking on the road as it ran past the yard fence to Austin. Rose vines overhung the gallery. A patterned coverlet of gypsy bloom lay between yard gate and front steps. Each time Dock or one of his cronies discarded a pocket flask Sallie gathered it up to save until she had enough to edge another flowerbed into which she poured leaf mold from the woods along the creek and droppings from her hen house. Her daughters called their home "Rose Cottage."[1]

In the spring of 1885, when five-year-old Eugene watched Ocie Cato boss three thousand longhorns up the trail past the yard gate, he spotted the gleam of oilskin slickers tied behind the saddles of every cowboy in the outfit, including his cousins, Tom and Baldy Alford. Longhorns were old hat, but slickers were something new and dramatic—unrolled, they could be flaunted like a bull fighter's cape to stave off an oncoming stampede![2] The idea would serve for games with the Black boys and was much more exciting than hearing at the supper table how a flood coming down Onion Creek had cut his Uncle Ocie off from his house on the Driskill Ranch, and how, even though he held the herd over two days, he had been unable to get back to tell his family goodbye. The family then included, besides Julie and two-year-old Ethel, Jennie Alford, who had come to live with them after her grandfather's death. A second child was on the way. Added to the rest of the day's news was Ocie's purchase of the 125 acre farm across

1. Rose Cottage: used in the Giles sisters' correspondence.
2. Accounts of this use of slickers in *The Trail Drivers of Texas.*

from Rose Cottage where he would move the family when he returned from the northern drive.[3]

By the time he came back from the Black Hills of Dakota via Buffalo Gap he and Julie had a second daughter, Myrtle Jane, and Austin was throbbing with plans for a grand new capitol rising where the cornerstone had been laid at the head of Congress Avenue in 1882. It would be of pink granite from up the river at Burnet and it would rival the white marble capitol in Washington, D.C.! Moreover, the hotel the cattle baron, Jesse Lincoln Driskill, was building on Pecan (now 6th Street) "so cattlemen could have a decent bed to sleep in this side of Chicago," would be open in less than a year.[4] Theodora Hemphill (that daughter, it was said, of John Hemphill, the jurist who had superseded Sam Houston in Congress) was already established on 2nd Street as madam of The Mahogany Hall, a parlor house befitting in grandeur both hotel and capitol. Her place had its name, not from the eight sets of polished mahogany bedroom furniture and the square piano in the living room, but from the complexional tincture of lovelies she imported from New Orleans to entertain her guests.

No real beauty herself, Theodora still was one of the sights of the town, taking the air up and down the Avenue with her long-haired, exquisitely groomed orange and white cat on the seat beside her. She was said to have come to Austin from New Orleans where presumably she had learned the tricks of her trade inasmuch as the Crescent City was known to be the prostitution capital of the nation.[5]

---

3. Deed Book 86, p. 182, gives Ocie's date of purchase from Henry and Elizabeth Neans as September 10, 1884.

4. Writer's interviews concerning Driskill Hall, 1924.

5. Theodora is listed in *Austin City Directory,* 1886-1896, first on Live Oak (West 2nd) and then on West 4th. See fn. 6, Chapter Five, "The Distant Tocsin." A full report of the Hemphill estate settlement, following the Civil War, is in Travis County Court Records. After payment of five exslaves' claims and other claims, Theodora received $1900 in gold. Under the then prevailing Texas constitition (1866) she claimed legitimacy declared for all children whose mother's slave status banned legal marriage. (The legitimacy clause was not included in the 1876 constitution.) After the court suit, Theodora vanished from Austin and her name was unheard until establishment of the house on 2nd Street.

Book 45, *Chattel Mortgages,* Travis County Courthouse: Entry for October 25, 1884, shows Goggan Bros. holding a lien on a J and C Fischer Square Piano No. 44273, "with stool and cover," for $300.

Description of the carriage-riding cat by James Grizzard, retired from the Austin police force, who remembered Theodora well. Interview by letter, *ca.* October, 1940.

Baldy Alford, his trail days ended at age twenty-one, deciding his brother, Charles, would have a better start with his Uncle Jim Maxwell (who was thinking of moving into town) than on the Giles' cotton farm, proposed to "go it on his own" if the Maxwells would take Charlie with them in his place. He would try starting a newsstand and

---

In the hey-day of the capital on the Colorado, Theodora's was not the only parlor for entertainment of the town's gentlemen; and she, though the only one of color, was neither the only nor the first demimondaine to parade in style down Congress Avenue. Jody's story of a predecessor fascinated the writer as a teen-ager. It seemed that descriptions of affairs at the mansion on Onion Creek when Judge Sneed was master and Mirinda mistress of the plantation made up much of the times' small talk.

Like any lady in her position, Mirinda was aware of whatever was going on on her premises. How she always knew is one of those subtle nuances of plantation life incomprehensible now when the grapevine is less important as a means of communication.

Mirinda knew, one peaceful morning, that a stranger in a plumed hat, holding her voluminous skirts carefully clear of gravel and dust, was coming up the quarter-mile drive from the Bluff Springs Road where the stagecoach plied between Austin and San Antonio. Met at the door by a serving woman who asked her into the drawing room and Mirinda's presence, the unexpected visitor explained, a little loftily, to her country hostess that she had heard so much of the place, she was begun to wonder if what she heard were real—or an exaggeration; she was from elsewhere, and had never been on a plantation.

The serving woman was bidden to see her refreshed in the dining hall, then to take her on a tour of the house, including the ballroom on the third floor. Down again, they would go through the dining hall and out to the back premises where she certainly must want to see how, when judgement had been determined, those found guilty were punished for unbecoming conduct.

Past the back door, geese and ducks wandered about, chickens dusted themselves in the sun and two Black men stood immobile, one on either side of what might have been a chopping block. A third man, equally silent and still, except for fingering coils of a blacksnake whip, leaned against a tree. Their eyes were fixed on the two coming through the door. If the stranger drew back a little from this theatrically waiting scene, she was propelled forward by the hand tightening on her arm till—her skirts fluttering wildly, the plumes on her hat dipping over the chopping block, the geese honking in alarm—a scatteration of chickens echoed her own startled squawks as the whip rose and fell! Once, again! Not too many times, not too hard.

With the same suddeness she was upended and hustled round the house, back to the driveway, unheeding a farewell from the lady left standing on the steps: "Now you have seen everything, may your visit prove profitable to you."

News of such an incident travels fast among those to whom it is of

tobacco shop at 119 East Pecan (6th) Street.[6]

In 1888, six years after the capitol's cornerstone was laid, less than a year after Frederic's death, and at the height of Theodora's reign over the local underworld, the glory of the new capitol building was flung open to the world. More appealing to Ocie than the state ball, the parades and all the rest of the fanfare on May 16th, was news that the Capitol Freehold Syndicate that had provided the wherewithal for the mighty showpiece was laying plans to extend its "world's greatest cattle ranch" from the Texas Panhandle to the country's "great Northwest"—namely to Yellowstone County in Montana.[7]

This was the year Alice would finish at Gilliland Academy with a gold medal for scholarship. It was also the year Eugene's formal education would begin in his father's granary with Jennie Alford the teacher. (Jennie had received a diploma at the Huntsville Normal School the year Frederic died.) Eugene's classmates were his cousins, his cousins' cousins (Mary and Buddy Thurman), and a few neighbor children. Ethel Cato, then five, was the youngest pupil, Andrew Payton, fifteen, the oldest. Maggie attended, too, but not Flora.[8] It was a convenient arrangement with Jennie needing to walk only over a bit of pasture, and down a lane with little Ethel in tow.

With money continuing so scarce, Giles young folk had picked a little cotton and Dock had paid them along with the familiar Blacks so they could have pocket change; it was something even Flora and Maggie could do.[9] But now boys in the communities were growing up and calling on Alice and Laura and Mollie—sitting through the evenings on the gallery with its weight of roses, taking them buggy-riding on Sunday afternoons, playing games like Skip-to-my-Lou, my Darling! at parties lighted by torches—cattail rushes soaked in

---

concern. It is to be doubted if ever again a denizen of Austin's demimonde presumed to call on the mistress of a local plantation.

6. See fn. 15, Chapter Seven, "The Hunter's Moon."

7. The Syndicate's Texas property became known in 1884 as the XIT Ranch for "ten counties in Texas." (See fn. 9, Interlude Minor.") According to the *Texas Almanac,* 1845-1846, the ten counties covered were Bailey, Castro, Cochran (part of), Dallam, Deaf Smith, Hartley, Hockley, Lamb, Oldham and Parmer.

The writer as a child believed the entire structure to be of granite and was somewhat let down on climbing the many spiral stairs to the top of the dome to learn that it was of painted metal coated over with granitic sand which provided an illusion lacking in modern paint jobs.

8. Ledger No. 9, 1937, "School Days."

9. Payments itemized in Dock Giles' small account books.

kerosene—flaring in an open space beyond the yard limit.[10] Perforce, the household routine would change.

It was noticeable that it was to Laura that a young Manor cadet, E. R. ____, when home from military school, always skipped and that intervals between his visits (and his letters) grew shorter and shorter, though usually he was accompanied by a friend the girls called Mr. V____.[11]

However pleasant the evenings, it was not too pleasant for these young gentlemen going back to Manor through The Field's dark turnrows where they were sometimes lost at midnight and not infrequently had their hair lifted by ear-splitting screeches of startled owls. After one such harrowing experience, E. R____ declared to Laura that Rose Cottage, though delightful, merited "Nowhere" as its proper address and thereafter included it on all communications to her.

Times were changing for Eugene as well as for the girls. When that summer was gone, having reached his eighth year and having been initiated into class routine in the granary, he was allowed to attend the newfangled "public school" a couple of miles up the road at Dessau. The teacher for the year was Miss Imogene Boyce, an erstwhile classmate of his sisters and his cousin Mollie at the Academy. He did not ride his pony to school, but walked with other children and, on some days, crawled—or, as they put it, "cooned it," creeping on all fours atop the rail fence to avoid the quagmire of the road in wet weather.[12]

The greatest change for everybody came the following spring (1889) when Ocie Cato sold his acreage and removed his family to Nocona, a new railroad town in North Texas. He was back in the employ of the Driskill-Day connection (Jesse Lincoln Driskill's wife was Jane Day, sister to "Doc" Day, a fellow cattle baron.)[13] This move put him almost at Red River Station, the old crossing place where the sprangle of Texas trails pointed into the main stretch of the Great Northern Trail (Chisholm Trail) leading to paying markets.[14]

---

10. Skip-to-my-Lou: a round game described by Mrs. Mary Thomas Simons, interview, 1977. The Memoirs describe the torches.

11. Letters in Laura's files.

12. Ledger No. 9, 1937, "School Days."

13. Deed Book 86, p. 162, gives the date of sale to J. S. Momsen as March 27, 1889. Sales price $5,000: $4,400 cash with $600 note at 10% interest.

14. Typescript of "Brief," J. Frank Dobie (found folded in Eugene's copy of *The Trail Drivers of Texas*) which says, "The Texans who went up the trail in the late sixties and seventies crossed their cattle at Red River Station, in Montague County, to point into the trace laid out by Jesse Chisholm."

It was then that he told his friend, John Driskill, anent the birth of a third child, "Whether this one is a boy or a girl, John, it is going to be named after you!" Much as they cherished their daughters, in their hearts he and Julie yearned for a son.[15] According to family lore, Ocie, aged fourteen, after a tiff with his stepbrothers, the Holman boys, had been sitting alone except for his horse under a tree on Barton Creek (a tributary of the Colorado, just above Austin) when John Driskill—Jesse's son—found him, took him home with him and started him at a job that was to determine his life style.[16]

On August 12th, another daughter was duly christened "John Ozella" and, although she was to grow up to be the family beauty, would never be known as other than "John"—not "Johnnie," or any feminine diminutive. Ten days later, back on upper Walnut, Daniel Rufus Payton died. Amanda, with her eldest, John Wesley, two years married to Lura Selma Lantz and her second, Ephraim Frederic, to Margaret Fulkes on Big Walnut, was left with a considerable estate; two teen-agers, Andrew and James; twelve-year-old Eula; Dora, seven; and Bradford Norman (named for her and Sallie), four. She would remain in their mountain home, a lady landlord, to become as

15. Mary Cato Williamson Swayne, Miles City, Montana (Myrtle Cato's daughter) gave this answer when asked about her aunt's name.

16. Interview with Ethel Cato's son, Edward Cato Tarbutton, San Marcos, Texas, summer, 1977. The writer recalls hearing a conversation between J. Frank Dobie and Walter Prescott Webb, in which it was said, "O. C. Cato is the next cattleman about whom a book will be written."

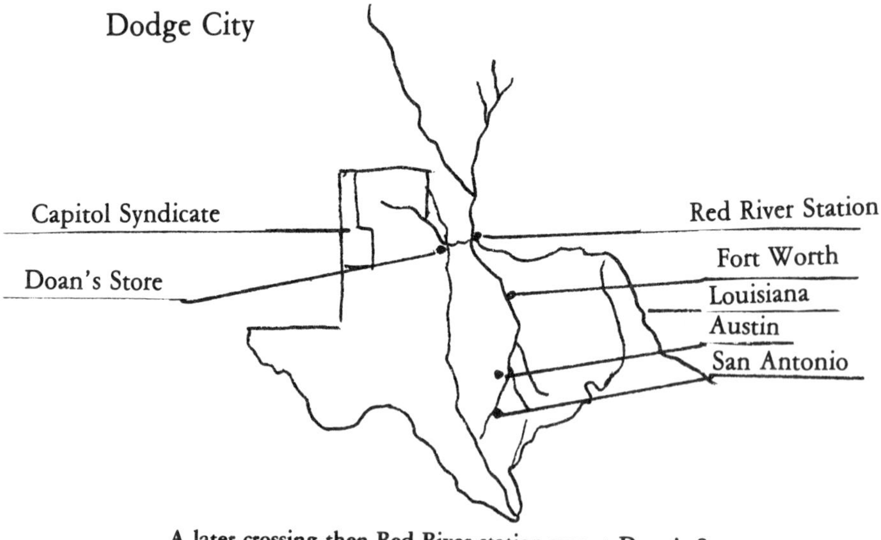

A later crossing then Red River station was at Doan's Store.

respected in business dealings as were her brothers and brother-in-law.[17]

In Austin, Baldy had hoped for a time that he might become a doctor and had enrolled at the University of Texas, then holding classes in the temporary Capitol (replacement of the limestone building burned in 1881) but after trying his hand at several different ways of making a living, one of which was selling meat for his Uncle John, he closed his living quarters over 410 East 6th and boarded a train for New Orleans to seek his fortune under more propitious circumstances.[18]

The Bible record for the 80s is a long one, including seven marriages, ten births of grandchildren, and nine deaths. It was the deaths of Frederic and Harriet that brought about for their children a shift in the focus of family life from the parental hearth to their own firesides; but at the same time an overwhelming force, culturally speaking, shifted them and everyone else from a pattern of out-lying *haciendas* (as in Mexico) to a pattern, English-manor fashion, of communities, each community revolving around a nucleus of country store-with-post office and the whole round the capitol at Austin. This over-all exchange was wrought by breaking the prairie sod for cotton and a consequent influx of tenant population.

The new order, while taking away much of the charm of the past such as supplanting open hearth cooking with fancy kitchen ranges (thereby bringing the hitherto outdoor kitchen under the main roof) and replacing silvery rail fences with the horrors of barbed wire, provided an opportunity for cultivation of the arts and for more frequent fraternization with one's friends and neighbors. Nowhere were such opportunities more fully recognized and utilized than by the ladies of Rose Cottage.

---

17. Amanda's funeral service at 2 p.m., Sunday, February 7th, 1932, was the largest this writer has attended. It was a beautiful day, the first "good day" after a stretch of winter. Amanda, a month short of 87, was the last known resident of Austin to have been born in the Republic of Texas. The funeral director was hesitant to add a hundred chairs to the three hundred customary in his chapel, but they were put in and quickly filled. So was standing room, overflowing into the hall and out onto the porch. Two representatives from the Daughters of the Republic, Mrs. Starkey Duncan, the president, and Mrs. Mattie Austin Hatcher, the University of Texas archivist, found a place on the portico of the Methodist Church across the street.

Looking back from a family car as the funeral procession approached Fiskville, one could not see its end. The unit of cars (someone later said 20) bearing the faculty of the State School for the Blind, where Miss Eula,

Amanda's older daughter, had long been on the staff, thinking to avoid the tedium of driving the slow seven miles in procession, had cut around and was then on another route in another procession they assumed to be Mrs. Payton's. As time dragged and miles lengthened with no way for occupants of the cars to make inquiry, it was late when they found themselves something like thirty miles north of Fiskville. They were close to Hutto beside the grave of a totally, so far as they were concerned, unknown.

Looking forward from the family car, one counted 50 vehicles of country folk, who had not attended the service in town, waiting at the Pioneer Cemetery (on a road off Rundberg Lane) for Amanda's interment beside Daniel Rufus and the two sons they had lost years before, one in boyhood, one in infancy.

All this for an unobtrusive pioneer gentlewoman whom Frederic told, when she asked him, "Why is my name "Amanda Bradford?"—"I called you that for a little girl I used to know," and one of whom her widespread bustling family (she had two dozen grandchildren, many of them married with children of their own) thought, as did her daughter, Eula, who first proposed to have the service in the Hyde Park cottage they had for some years occupied together: "Mother was just a quiet home body." (The writer's choice for her favorite aunt would have been the Central Christian Church where Dr. Henry Clay Garrison, who delivered the service, was pastor.)

18. *Austin City Directory,* 1885-1889. A letter of reminiscence, Baldy to Eugene, December 27, 1839. Reminiscence to the writer, summer of 1914.

It was to become one of the family love stories that the disappointed girl he left behind, Nealie Morris, would wait for him until, in their middle years, he returned for her after Katie, the girl he married in New Orleans, died of breast cancer in 1916. To satisfy her anxious mother—and to the displeasure of her husband and daughter—Katie would be buried a Catholic long after she had withdrawn from the Church's rituals.

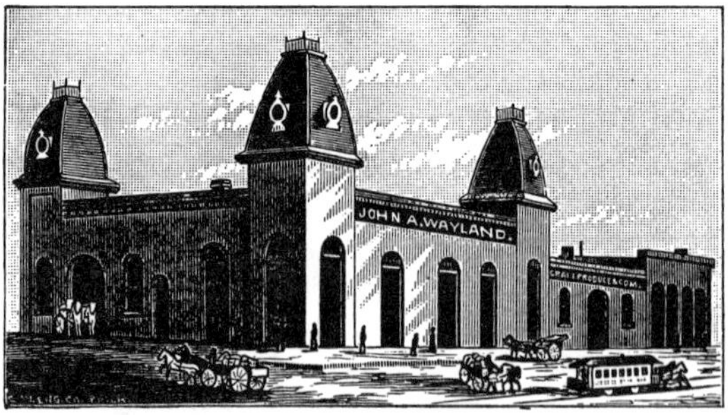

Corner of North Avenue (W. Fifteenth) and Lavaca Street.

JOHN A. WAYLAND, Produce House, Austin.

AMANDA-1927

ZACKIE—1925

END OF THE COMANCHE TRAIL IN TEXAS

Cynthia Ann, the captive maid,
By her blue, blue eyes betrayed,
Was taken from her tepee floor
To her father's alien door.

There just a little while she stayed
As round her knees the Flower played;
Looking in that alien fire
She brooded deep on old desire,

For, though she knew Nocona slain,
And Pecos, too, she knew
One son yet ranged the hill and plain.

The little Flower drooped and died—
Or so the aliens said—they wept
To take it from the close hearth side
And lay it in the prairie wide;

No Pecos here, no Quanah near, Nocona gone—
Why linger by that cold hearth stone,
Longing, weary and alone?

So Cynthia's own wild spirit fled
To join Nocona's spirit! Dead?
That's only what the aliens said.

—M. Atkinson

Interludes

during the attempted change-over from hacienda culture to plantation culture and from cowhides to "vegetable wool" as a principal product when philosophic tobacco chewers opined the human heart could not withstand speeds of more than fifteen miles an hour on the railroads rumored to be approaching Texas—the heart would simply burst.

Interlude (Major)

Then came the terrible farmers;
Nothing of play they knew, only,
They caught our fathers
And set them to labor, too.
—Rudyard Kipling, "Legends of Evil"

The shift from cattle to cotton meant a shift from saddle to plow for Black Land men; it also meant their being circumscribed in space. For women it meant conveniences such as sewing machines and Sallie's Comfort Range to lighten their labor, and luxuries to brighten their days, like her parlor organ, for cotton was "the cash crop."

However, in the thirty-year interlude between the Civil War's close and the Giles tragedies of 1897, that period when men in ten gallon hats with bandannas about their necks superseded men in gray as figures of romance, two inventions wrought a difference in life style and landscape that would have marked the end of the pioneer era on Walnut, even without cotton. One was the Winchester rifle, the other, barbed wire.[1] Yet it was not, looked at in cultural perspective, the actual repeating rifle that overshot the Indian's bow by a couple of miles, but an intangible culture lag that cleared the Comanches from the High Plains, together with the buffalo. Had the confederated *Comanchería* that extended from the Saskatchewan in Canada down into the Texas Hill Country possessed coined money or other form of specie with which to buy the rifle for themselves, it becomes a question as to whether Polk's idea of Manifest Destiny might not have died a-borning and the United States have been restricted to east of the Mississippi, except for Texas and Louisiana.[2]

Suffice it to say that by the time Ocie Cato began going up the trail with beef on the hoof for northern markets, he rode over an earth

---

1. *Webster's Unabridged Dictionary* says the Winchester, first highpower repeating rifle, went on the market right after the Civil War. Barbed wire, according to the *Texas Almanac*, 1945-46, was invented in 1873.

2. That annexation of the Texas Republic was then so unpopular it had to be by contract, requiring only a majority vote of the Senate, instead of by treaty which would have required a two-thirds vote, is something citizens from other states rarely understand.

Old Timers say silent arrows were much more effective than the musket; that it was only when two or three men could stand behind a barrier picking off Indians before they came within bow-shot that the buffalo were destroyed.

so littered with skeletal remains of the buffalo on which Plains economy had depended for sustenance and trade that a man urging his horse beyond a trot risked its breaking its leg and his neck, or both their necks, if its feet became entangled in the bones. Eventually other scavengers would come to gather up these relics left by human and avian vultures, load them into wagons, haul them to the railroad and ship them east to be ground into fertilizer.[3] The last living remnant of the shaggy herds that had roamed the Plains by millions would remain on the Goodnight homestead in the Palo Duro Canyon of the Red River where Mrs. Ann Dyer Goodnight refused to see the last bull and cow slaughtered for their hides and tongues.[4] In the extremity thus created, Comanches, erstwhile lords of these great herds and this vast wasteland, rather than starve, entered the federal reservation above Red River, led by Nocona's son, Quanah Parker.

Down on Walnut, by 1897, the log cabin, as a dwelling, had gone the way of the buffalo and the Indian. For the most part, remaining log structures were corn cribs, shedded at the side for cattle stalls like the one Sallie moved up from her father's "old place" and the barn Jim and Jane had on their place near Sprinkle.[5] Almost as rapidly disappearing was the split rail fence. Tales of how Martin Wieland, one of Walnut's earliest settlers, brought enough cedar from The Mountains to split-rail fence a square mile (a section, or 640 acres) and how Mrs. Fritz Kruger was the first person out with a yoke of oxen and a plow "busting the sod" to throw up a furrow and halt a prairie fire threatening community fencing, along with stories of Old Lil, the Giles' longhorn who gave blue milk and was so smart she could budge any rail to let herself out of a cowpen, were become legendary—something to tell children.[6] Barbed wire strung between cedar posts ten to twelve feet apart, in stilling the woodcutter's axe, was proving to be as much a labor-saving device for men as sewing machines and ranges for women.[7]

---

3. See Appendix H for trail riders' experiences.

4. Tongues were preserved by drying.

5. The two-crib, with dogtrot, log structure Frederic built and Sallie moved up to her home has been returned to what is now the Jourdan-Bachman Park by courtesy of Carl and Mary Jo Anderson (see fn. 2, Chapter 3). The Anderson Development Company presently owns the Giles farm land.

6. "Blue" milk or "blue john" was called after a bluish stone worked in Derbyshire, England. In the 1942 Ledger, Eugene comments that Old Lil gave only two or three pints of this lacking-in-butterfat milk a day and that it took a pen full of longhorns to supply the dairy needs of an average family.

7. See Appendix G for Eugene's research on the types and building of rail fences.

At the same time that fencing-made-easy allowed a man to enclose his property with the labor of a few hours, it ended his freedom to travel across country at will.[8] A man not only fenced his fields and his cattle pens, he fenced his pastures as well. A reservation Indian, leading an innocent hunting party down into Texas, epitomized the situation when, lost in this new entrapment for men and animals, he flung up his hands, crying in despair, *"Alambre! Alambre! Todas veces, alambre!"* (Wire! Wire! Everywhere wire!")[9]

The ring of the rail splitter's axe was not the only sound silenced on the Black Land by 1897. Howl of the wolf and laughter of the coyote were no longer heard at close hand; nor was the drumming of prairie chickens, echoed by the swelling boom! of a strutting turkey gobbler answered by shrill gibble-gibbles of his hens. Gone, too, were the deer, up into the hills or down into the river bottoms, and the fish from the swimming holes on Walnut. Eugene and his cronies had never seen anything there but minnows. John Lafayette Alford would write him years later of flights of passenger pigeons darkening the sky in their migrations—flights to be reduced finally to a single bird, stuffed, on a stand in a museum.[10]

It helps one to understand how it came about that when Frederic and Harriet's younger children and older grandchildren were grown, the pioneer era was already past and gone, if one will read the map of Texas in more ways than geographically. To begin with, its shape is an eccentric trapezoid. The top of its Panhandle lies just below the midline of the United States. Its tip touches the Gulf of Mexico. Rainbowed round its middle, and so around the Hill Country, which is North Central Texas, is the Balcones Faultscarp running roughly from Gainesville, in the Red River Country of North Texas (though less conspicuous in its upper stretch) to Del Rio on the Mexican Border. On the Mexican side, people call the Rio Grande, at this point, the *Rio Bravo del Norte.*[11]

---

8. "The open range ended in the 1880s and 1890s with the advent of fencing" (due to barbed wire), *Texas Almanac,* 1845-46.

"After barbed wire was invented, they tore their rail fences out and used their rails for firewood." Excerpt from the J. L. Alford letter to Eugene, November 11, 1939.

9. The writer did not set down the reference and does not recall in which item of Texana this incident is recorded.

10. Alfred letter. See fn. 11, Chapter Nine, "Miss Sallie's Boy."

11. This is quickly apparent from public signs. Original Spanish explorers called the river at its mouth, *El Rio de las Palmas* for the abundant growth of a tropical plant said to be native in the United States only at the tip of Texas and

The Nueces is an older folk boundary than the Rio Grande. Tracing back up from the Gulf, keeping to the easternmost branch of the Nueces, one rounds the Brush Country, labeled on early maps the "Mustang Desert" or the "Wild Horse Prairie." Here only Spanish was spoken until the irrigation project just before World War I brought a rush of northern grapefruit growers and truck gardeners to try their luck in the Magic Valley, new name for the Rio Grande Delta. Until the Battle of San Jacinto when Sam Houston put the cabitz on Santa Anna, the Mexican states of Tamaulipas, Nuevo Leon and Coahuila reached the Nueces and so came almost to San Antonio. The Spanish language, like the longhorns and the mustangs, has remained on the land.[12] Today, barring orchards and gardens of the Valley and the old towns along the Rio Grande, this is an area of mesquite and prickly pear, of javelina, rattlesnakes, dust devils, oilrigs and the King Ranch, where a stranger, defying the barbed wire boundary, can disappear into the brush and never be heard of again.[13] It has been known to happen. All this semi-wild country with an imaginary boundary straight down from the eastern line of the Panhandle to the Nueces, then winding with that river to Corpus Christi, is West Texas. Tracing up the Rio Grande, true mountains appear—an extension of the Rockies overshadowing the Big Bend—an area which is still The Frontier.

Eastward, above the Scarp, the Hill Country, or North Central Texas, was largely settled by German families Sam Houston brought in to serve as a buffer between the Comanchería and Anglos moving up from South Texas plantations to replace the Tonkawa on the black prairies of South Central Texas. The Comanches, rejected when they would have allied with the Texas Republic, were then finding the Scarp's limestone balconies convenient ramparts from which to pot-shot congressmen on the streets of Austin.[14] By 1897, Germans of the

---

the tip of Florida. See fn. 6, Chapter 6, "Good Man Gone Wrong," for "Dock's" letter from "The Rancho de la Palmettos."

12. A map reproduced in the Archives Division of the Texas State Library in 1965 shows the original extent of North Mexican states.

13. Javelina: native wild pig. The writer's earliest acquaintance with this animal was with the mascot of a fraternity house on the way to Woolridge School. The acquaintance ended when the creature quite literally upset one of the ladies of the town in her own yard and was banned from city streets.

Dust devil: whirlwind common to arid and semi-arid areas.

14. That a law is still on the books forbidding state monies to be sent out of the city after sundown lest there be a Comanche attack is given as the reason for the state's out-of-town employees' paychecks not arriving till Tuesday if they fall due on a weekend!

Hill Country (called Forty Eighters to distinguish them from Central Europeans in older settlements and from 1870ers who would come later), finding themselves at home with the wool and mohair industry that extended and still extends across northern Mexico, dotted the alkaline soil with gabled stone houses and scored it with their sheep and goat runs.[15] Until World War I most of the German settlements clung to the German language and taught it to their children as their native tongue, wherein they were supported by a law of the Republic of Texas requiring state papers to be published in English, Spanish and German.[16]

Passing immediately east of Houston and straight as the crow flies from the Gulf until it curves east just before reaching the Red River, is the Timberline. It delimits a Bush Whackers' paradise of tangled thickets and the pine lands—the lumber country—that extends across the Sabine into Cajun and Creole (French-speaking) Louisiana.[17] This is East Texas and Deep East Texas. Going back along the Red from the northeast corner of the trapezoid, Texarkana, to the angle of the Panhandle, is a fringe of Anglo counties called North Texas—a spill-over from the United States. This fringe separates the Hill Country from Oklahoma which was, until 1907, the Indian Territory where the Comanche-Kiowa were congregated by the federal government after annihilation of the buffalo.

To indulge in fancy, the trapezoid is a little like some antidiluvian creature emerging from the mists of time with the tip of its tail still splashing the sea. Delving into Texas' past, the thought is not inapt, for, reading the map this way, it is easy to see how Old Texas, culturally and linguistically and topographically encapsulated by alien terrain and alien mores, had begun to develop a character peculiarly its own before the first fleecy seed of cotton was dropped into black prairie humus—even in Indian time, long before the first Spanish

---

15. Forty-Eighters: from the year of immigration. These colonials differed from Old Texas Germans in that they had not been, like the latter, citizens of the Republic of Texas; that they were ranchers and not planters; and were Unionists. *Cf.* fn. 27, Chapter Four, "The Distant Tocsin."

16. *Gammels' Laws of Texas.*

17. "Cajuns"—folk use of "Acadians": French or French-Indian people expelled by the British from Nova Scotia in 1755. Creoles: originally pure bloods born to Spanish parents in the Americas. "Creole" is a French spelling of *criol'*, shortened from *criolla,* which is derived from *criar de los colonias.* Today it has many ethnic applications. Atkinson, *Creolism and Racism on the Gulf Coast,* 1977.

cow, sheep or goat was driven across the Rio Grande.[18] Then reversing direction, it is equally easy to follow the Anglo frontier from East and South Texas past San Antonio and down into the Brush Country, out into the mountains of the Big Bend or up onto the Plains, and to understand how (rephrasing the traveler in *The Lilies of the Field*), when one has crossed the Mississippi, "South" ceases to be a noun and becomes an adjective modifying "West." It is also easy to understand, considering the rapidity of this movement, how soon the pioneer era was to be forgotten.

---

18. See Appendix I on ethnicity.

FLORA AND MAGGIE BELLE

Interlude (Minor)

Too exquisitely fragile
And delicately fair,
It did not bear handling,
My castle of air!

Ah, it was lovely,
A-light and a-gleam,
But then, fell a shadow—
And vanished! my dream.

—M. Atkinson

John Ozella, third Cato daughter, was born into a social suspension between Austin, where her parents' roots were entwined, and Miles City (Montana), which would soon be their home. Her mother, her Cousin Mollie, and her little sisters, Ethel and Myrtle, brought her down to Walnut for an introduction to and an extended visit with her kinsfolk when she was a year old.

The splinter-fire new town of her birth, Nocona, in North Texas, was but two years old, having come into being (1887) due to an extension of the Missouri, Kansas and Texas Railroad. It had about it a stir and a bustle quite unlike the community centers on Walnut and Gilliland, including Manor, which also was railroad born.

Ambitious young men, single, but hopeful of acquiring both a family and an estate, were appearing in Nocona from all directions: neighboring Belcherville, a German community; Red River Station, the cross-over point for cattlemen taking their herds on the hoof up to Jesse Chisholm's trail that led through the Indian Territory to northern markets; and Old Spanish Fort, another crossing farther east, used, even before the Spanish got into Texas, by *couriers de bois* (''runners of the wood'' - French pack traders) carrying contraband to Indians along both sides of the *Riviere Rouge.*[1] Romantically, the

---

1. A telephone conference May 6, 1978, with Mrs. Sam King (born Annie Thurston, 1882) provided much information on Montague County. Hers was among the families moving into Nocona from Red River Station.

Among families moving down from Old Spanish Fort was that of Joe A. Justin, whose hand-made boots and shoes are memorialized by the Nocona Boot Company (controlled by his daughter, Enid) and the Justin Boot Company in Fort Worth (under management of his sons). The Justin name is now known in connection with cowboy boots throughout the United States.

town bore the name of Peta-Nocona, who, with his son, Pecos, died in the Battle of Pease River in 1860.[2] His other son, Quanah, escaped to replace him in the chieftainship, but the mother of his sons, Cynthia Ann Parker, was captured by United States troops, only to pine away and die when returned by federal authority to what God-fearing folk of that day looked on as "civilization."[3]

On the morning of November 15, 1890, Laura Giles, pretty and a year past twenty, on the way home with her aunt and cousins, was in charge of their tickets; of Charlie Alford's gift of candies, fruits, and other foodstuffs from his well-stocked grocery shelves—and of young John Jourdan, who was well on in his eleventh year. "Never," she wrote Alice next day, "have I ever seen a child as happy as John when we left Austin." Wont to climb aboard a chair on the gallery at home to watch as long as he could follow a plume of train smoke across the prairie toward Manor, the boy would be satisfied—at least for awhile—only now, instead of going east on the Houston and Texas Central, the International and Great Northern would take him toward the new frontier between the United States and a mysterious, unknown Indian Territory.

Laura, too, had reason for rejoicing over a change of scenes. It had been a sad year for The Field. Jody's sister Fannie, failing to recover from the birth of her second son and seventh child, had been at the point of death in February when Eugene and John were celebrating their tenth birthdays. Neighbor ladies, two at a time, had taken turns sitting up with her through the nights—among them Sallie Giles and

---

2. The Battle of Pease River (1860) in the Texas Panhandle was the last formal Comanche struggle for any part of Texas. See fn. 11, Chapter Four, "The Distant Tocsin." In this battle Peta Nocona, chief of the Comanches, was reported to have been killed.

Mildred P. Mayhall, ***Indian Wars in Texas,*** using information not revealed by "Noconi's" family until 1934, says he did not die, as supposed, in the Pease River engagement—but from an infection about 1862—that his Mexican servant was killed and the body mistaken for his, hence the military report to that effect.

3. Cynthia Ann Parker, according to historical references, was captured when nine years old, with her brother John, aged seven, in a Comanche attack on Parker's Fort in Limestone County, 1836. Both children were brought up as Indians, Cynthia Ann to become the wife of Peta Nocona, John the husband of a girl captive from Mexico who is said to have effected his turning to the cattle industry on her family ***hacienda*** where his nephew, Quanah, eventually learned it and taught it to his people after the slaughter of the buffalo. Cynthia was pregnant when captured. The child, Prairie Flower, died shortly before her mother.

Mrs. Laura Yager.[4] She died on the 18th (presumably of what was called "childbed fever") leaving her baby, Ernest Flynn, to Jody's care. The ladies of the community were greatly exercised to have in their midst a man struggling to fend for a womanless houseful of small, stair-step children with the eldest, Mary, barely ten; after her, Augustina, then William (called "Buddie"), who was the same age as John and Jody's William; then Josephine, Helen and Delia. Those most concerned thought of placing the children in private homes, or perhaps an orphanage.

They should have been warned when one of the little girls told Jody, "Papa is going to take us away as soon as the baby is old enough to leave"; but who is less apt to note a differing point of view than someone bent on doing good? So it came about that when, after three months of attempted baby diets, little Ernest was buried beside his mother in the plantation cemetery, Flynn Garrison Thurman vanished overnight with his wagonload of little girls and his seven-year-old boy. Whatever else might happen to his children, they were never to be separated from him, their father, who had loved their mother's little finger (he told Jody) more than his own life.

It would be long before they were heard of again and only Buddie, grown up, would return from time to time, resuming the family relationship.[5]

Added to this trouble and sorrow was concern for Sallie who was beginning to feel ill effects from a goiter on her neck. However, getting into Austin through mud and slop left by a torrential rain and seeing that everyone was ready for the train, erased all else from Laura's mind for the moment.

A greater surprise was waiting for her and Mollie thirty miles

---

4. Reference: Laura's letter addressed to Alice at her Aunt Jane's in Austin, February 13th, 1890; for deaths of Fannie and baby, family Bible.

Laura and Mayette Yager were newcomers from Virginia who bought the Ben Earnest place in December of 1887 from Z. P. Jourdan (Deed Book 79, p. 203) who had it from his father in 1881 (Deed Book 50, p. 338). Frederic bought it from the Earnests when they moved into Austin in 1874 (Deed Book 42, p. 865). The Yager family included the maternal grandparents, Mr. and Mrs. J. C. Bowman, several daughters and a son (perhaps two, for one died in childhood). They promptly affiliated with and became members in the Old Rock Church (by then Baptist, though it was begun in 1856 as a community project) at the Iron Bridge on upper Walnut.

5. Jody told the writer of the baby's death and the father's rebellion concerning his children, in the writer's childhood; Buddie's visits continued until word of his death in West Texas came when she was a teen-ager.

out—at Taylor, where they would shift from the I and GN system to the MK and T that had been so recently extended to Nocona. They were half an hour late getting in over a roadbed washed out in places, to find E. R__ and Mr. V__ waiting on horseback. The train huffed and puffed for an hour and twenty minutes, starting up four times, backing back three, the horsemen following each time till finally it speeded up and was gone, leaving them up the track a mile out of the way on their return to Bryan.[6] A letter to the girls would say it took them all night to make it to quarters before reveille.

A gossamer thread in love's young dream, this episode was the closing act in a boy-and-girl idyl during which photographs were exchanged, invitations to college affairs extended with military formality, cakes and bouquets sent over the rail to the college dorm. Sad that such should be the ficklesness of woman's fancy titillated by attentions from an exotic stranger!

Soon after leaving Taylor train-sickness overtook first one and then another of the travelers; but by the next evening, though they were more than two hours late, getting in at 10 p.m., the two young ladies were recuperated and ready to be on the go all the next day, escorted by Robert Clark, already Mollie's devoted admirer. Pharmacist and partner in a business that (according to the *Argus,* Nocona's four-page weekly news shett) sold everything from prescription drugs and patent medicines to hominy and grits, maple syrup, paints and wallpaper, young Mr. Clark made his interest in Mrs. Cato's niece manifest in every permissible way and on every possible occasion. Grandfather Giles would have approved of Nocona's weekly, which carried under its title a banner: THE ARGUS O'ER THE PEOPLE'S RIGHTS DOTH AN ETERNAL VIGIL KEEP. NO SOOTHING STRAINS OF MAIA'S SONS CAN LULL HIS HUNDRED EYES TO SLEEP.[7]

A pretty stranger like Laura in a frontier town full of footloose young men was guaranteed instant attention, the more so when Mr. Clark introduced his circle of friends at Nocona's social center, the Commercial Hotel. Mr. B__, the first to claim her favor, was (she wrote Alice), though charming, too short for her taste. She preferred escorts as tall or taller than herself. The next, Mr. M__, may have made up in height what he lacked in character, but she was blind to any fault and more than pleased to receive his invitation to a Grand

6. Letter from Laura to Alice the morning after the party's arrival in Nocona.

7. One falling-to-pieces copy of the *Argus,* dated January 29, 1891, is among the Giles Papers. Maia (May) the goddess of spring.

Thanksgiving Banquet and Ball to be given by the Nocona Social Club at the hotel on the evening of the 27th. Senior members of his family were listed among the sponsors.

Back on Walnut for Christmas, all Laura could talk about was Nocona and Nocona people. Her chatter annoyed her Aunt Jane and disturbed E. R__, the latter because too much of it eulogized Mr. M__, her aunt, on returning from a visit to the Catos, had been only too happy to clean Nocona's light-colored dust from her shoes—the area smacked too much of the range country for which she had little taste.[8]

Spring of 1891 brought Julie and Ocie's announcement of Mollie's approaching marriage to Mr. Robert Browne Clark on April 2nd at their home, a function followed almost at once by their removal to Miles City in the far Northwest where the cowboy, in whose keeping Jane had feared for the welfare of her baby sister, had accepted managership of the Montana Branch of the Capitol Syndicate's XIT Ranch—lauded as "the greatest cattle ranch in the world."[9]

For Ocie, the Interlude was over. As a farmer on Walnut, he had been like a fish out of water. Now he was returning to the saddle for good and all.

The following November found Laura again on Red River, visiting the Clarks and Jennie, who was mistress, in its second year, of Nocona's first public school.[10] This time the highlight of the month

---

8. Laura refutes Aunt Jane's opinion in her letter to Alice, November 25, 1890.

9. Ab Blocker says in *The Trail Drivers of Texas,* "In 1884 I drove a herd of 2,500 cows and heifers from Tom Green County to Buffalo Springs, in Dallam County for John Blocker and delivered them at the XIT ranch. Old Barbecue Campbell was in charge of this ranch, which was owned by a syndicate that had bought a lot of cattle down in South and Southwest Texas with which to stock it . . . Old Barbecue Campbell was undecided as to selecting a brand to be used by the ranch syndicate and I suggested XIT. It pleased him so well he decided to use that brand, and it became known all over Texas, Oklahoma and New Mexico as the XIT Ranch. I branded the first cow to carry the XIT brand." See fn. 7, Chapter Eleven, "Rose Cottage."

10. This second visit is an assumption based on a formal card with date similar to another kept as a souvenir of the Thanksgiving ball in 1890, on presence of the 1891 copy of the *Argus,* and on the fact that visits of the Giles sisters to their cousin in Nocona were fairly frequent through a number of years.

Jennie's diploma, recorded in registrar's office, is dated 1887. Family report is she was valedictorian of her class. Mrs. King remembers her as the school's principal and teacher after its second year, following Sherman Williams during the first year.

was the grand social ball given in Williams Hall by the Nocona Cornet Club, of which Mr. M__ was a member.

When the Manor cadet had questioned her all-too-evident interest in the cornet player, Laura's reply was that she had had her fortune told and for her the future was rosy. He wrote again, suspecting the worst, but wondering what it really was with Mr. M__, saying plaintively, "I had not thought, after we had been going together for four years, you would flirt with anyone else—and I know I wouldn't." Receiving no satisfactory response, he conceded the field to his suspected rival with a sadly gentle, "Goodbye, darling."

When she came home this time, in her mind everything was settled: Mr. M__ was coming down to Walnut to meet her parents. When he arrived not long thereafter, she gave no thought to nineteen-year-old Flora's presentation of her big doll, Stella Lillie; or to Maggie's welcome in what was to him unintelligible gibberish; or to their being seated at the family table. Sallie Giles' "afflicted children" had always been received with kindly courtesy wherever they were, because they were hers; and she had taken them everywhere. This guest felt differently. he could not be gone soon enough. That anyone should flee the sight of the two was a shock, both to Sallie's sense of decency and to her ego, however, she would consider the source and thank her heavenly stars Laura had found the man out before matters went any further.

Laura had made no secret of what she thought to be a conquest—the community would buzz for awhile over Mr. M__'s abrupt departure, but her entire family connection would deal with the talk. Aunts, uncles and cousins making light of whatever was said, shrugging that she was well rid of a man "with no prospects" (although he belonged to a North Texas family of good standing) found it easy to suggest that he had seized on an excuse to turn tail and run due to a sense of his own unworthiness.[11]

She would survive her disappointment with dignity, but also with an awakening to the transigency of youthful joys and an awareness of gossip going the rounds that the latest medical advices exonerated Mr. M__ for his flight—though he could have been less obvious about it for politeness' sake. Heredity was just beginning to be looked on as a factor equal with property in considering a marriage so that there were those who privately congratulated E. R__ that with him, too, matters had gone no further than they had.

---

11. Climax of the cadet affair is based on letters; climax of the Nocona affair is pieced together from reminiscences of Jody and others.

MOLLIE ALFORD

She put E. R__'s letters and photograph and the Nocona invitations and programs away in a box: mementos of a girlhood lost when she became aware of a shadow over her life. As spring merged into summer, one great glad tiding did come to brighten her spirit—the Catos had at last their longed for boy, and John Ozella and her sisters a baby brother—Leo Elmer.

LEO, The Beloved.

Gotta have a home!

PART THREE

The Cotton Towns

. . . cotton is a universal fabric; wherever people wear so much as a loin cloth cotton is sold.

—*The Story of Cotton* (brochure) Anderson, Clayton and Company of Houston, 1934.

Cotton, anthropologists say, made the world clean. Linen, just as washable but rarely grown, was reserved for the elite. Even for the elite, silks and woolens would bear little washing.

12

Sprinkle: Virginia Outpost

Bob Sherrill wrote once in a bit of doggerel in the Texas Observer something to the effect that ". . . out where the tobacco spit's a little spittier, the grits a little grittier, that's where the South begins."

Braxton Barr was in his first year as Sprinkle postmaster and proprietor of his grandfather's commissary when Ocie Cato bossed his last trail herd past Rose Cottage. Even before the post office sign appeared, Braxton's mother's cousin, Thomas Jefferson Rowzee, brought a new wife out from Virginia to share a home with him and his father, Old John, a little south of the commissary on the east bank of Walnut. They were just across the creek from his first wife's sisters, Jane Miller and Margaret Dunson, who had been the Tweedle girls of Hornsby's Bend. They had cared for their sister Julia's two-year-old daughter, Julia Ellen, since the moment of the child's birth and her mother's death.

Back in the Old Dominion Erasmus Sprinkle and John R. Rowzee's wives were twin sisters. As a consequence Tom Rowzee and Cremona Sprinkle Barr were first cousins. In Texas, the Rowzee men had been first at Hornsby's Bend with Erasmus's brother, John P. Sprinkle, then at Gregg's Corner where William Gregg's store looked straight down on the iron bridge at the Cameron Road crossing of Gilliland Creek, after which both the creek and a turn-off from the main road led past the Hill home and the old Cain place to Manor.[1] On the road just before the crossing was the home of the Gregg postmaster, C. C. Baker, whose first wife, Virginia Ellen Rowzee, had lived but three months after their marriage in 1878. The 1880 Census shows Tom Rowzee and his father with Baker at this time.

---

1. A concrete bulwark of the old bridge is visible just west of the present crossing.

It was but natural that to the Old Dominion Tom should return as a widower in search of Clara Vinita Marshall whom he had sought in the beginning to bring with him to Texas. He could not then persuade her to part from an ailing father.[2] When she arrived at Sprinkle, to be familiarly known as "Nish," the Millers and Dunsons begged to keep Julia Ellen, who seemed more theirs than her father's, and he let the child stay where she was.

Jim Maxwell's land stretched The Field to the north edge of Barr's holding. West of the creek, his uncle, Alexander Carson Maxwell, was settled with a large family of six daughters and four sons—hence, all in all, Sprinkle as a community bore witness to a powerful Virginia influence in its cultural origin.[3] When Jane and Jim, after building

---

2. Reference: Mabel Rowzee Penick (Mrs. E. Payson Pennick, Jr.), from family Bible. All Rowzee family records are from Mrs. Penick.

3. Maxwell family heads who came to Texas were Wallace, Alexander and Harvey. Wallace and Mary Corry Maxwell had two sons: Thomas and James.

Thomas, who was tubercular, married his cousin Virginia in 1872, and died in 1875. There were no children. Alexander Carson and Augusta Owens Maxwell had six sons: Fielding Pope, William Worth, Thomas Owens (Tom O.), Frank A., James Hardy, Jesse Worth, and four daughters: Florence Virginia (1-Mrs. Thomas Maxwell, 2-Mrs. R.J. Loving), Louisa Hadassah (Mrs. Alexander Campbell Hill), Ella ____ (Mrs. Thomas Washington), and Janie Alice (Mrs. Milton Morris). Harvey and Margaret Davis Maxwell settled further to westward.

Esther, 11th book of the Old Testament, *King James Version,* is the story of Hadassah, a young Jewish girl Ahasuērus (Ahas-u-ē-rus), overlord of 127 provinces between India and Ethiopia, took in place of Vashti, hitherto his queen. This came about when he, somewhat in his cups during a second day of banqueting lesser potentates of the realm, summoned Vashti to parade her elegance and charm before his court. She, feasting the royal ladies in the women's quarter, ignored the summons. What became of Vashti after that, we are not told.

Hadassah, as Queen Esther, hearing from Mordecai, her elder cousin and foster father, that Ahasuērus' favorite, Haman, had set their lord up for assassination, was rewarded for queering the deal by her husband's granting her request that Haman's ten sons hang with their father on a gibbet intended for Mordecai. She was further granted authority to issue for each of the 127 provinces an order to destroy all enemies of her people, including women and children. This act of genocide, excluding only those who took refuge in conversion to Judaism, amounted to the slaughter of 75,000 persons.

So blood-spattered a name as Hadassah hardly seems to fit a lady of gentle birth like Jim Maxwell's cousin Lou; nevertheless, every biblical word and name in that Victorian era was sacrosanct—as Holy Writ, its hidden

W.B. BARR'S Store, 2nd phase, 1893.
—Courtesy Mrs. Onie Jackson Kruger

and moving (out of the "yellow house") into a two-story house, later removed to Austin, Jim becoming a partner in the wholesale and retail business of JOHN A. WAYLAND-GRAIN PRODUCE AND COMPRESS (according to letters on the building) at the corner of Lavaca and North Avenue (15th Street), they left other Virginians—the E. Payson Penick family with several sons and daughter Mary—in the two-story house.

No Virginia feature was more in evidence than the tobacco habit. Tobacco would never be a Texas crop but Texas men would smoke and, more particularly, chew, with a gusto abominated by their wives on cold winter days when "ambure" missed its mark to sizzle on a hot stove lid.[4] Nevertheless, even among the elite, women might dip a little snuff in private, smoke an occasional corncob pipe or "take a chew" to ease an ailing tooth—an openly secret vice, as it were.

After seven years Brack Barr abandoned the one-room commissary to open a general merchandise store with a gallery on the front. To accommodate his burgeoning trade, he called on his cousin, William Maxwell Jackson, Fenton Sprinkle Jackson's son (Fenton was Cremona's sister) to bring his family out from Virginia and keep the store's books. By that time, Tom and Nish had a son, Montella, and twin daughters, Mabel and Irene.

The day of the grand opening (1893) found a black-draped tripod across the road from the store with a professional photographer's head under the cloth. Gathered on the gallery and in the road were forty or so men with as many of their teams and mounts as could be got in focus, George Barr's father-in-law's team of oxen to the front. George, married to Carrie Criswell, lived on the Criswell place across from Walnut, north of the new Cameron Road and south of the old Jourdan Crossing.

The resulting picture shows white-bearded Erasmus Sprinkle in the doorway of an offset to the new building (presumably the out-

---

meaning to be searched out and expounded by those enlightened through revelation. Following this thought, Hadassah or "Esther," stands forever an exemplar of daughterly, wifely and tribal fidelity in contrariety to Vashti, prototype of the Woman of Now, who would not forego her personal interests to pander to her inebriated husband's vanity.

4. Ambure: Texanism for *ambre,* French term for any amber colored liquid. The *Handbook of Texas,* 1952, says that, according to Justice of the Peace Minor, the community of Ambia in Lamar County received its name because the men were "the greatest tobacco spitters he had ever seen." (Tobacco spitting is today superceded, at least in Texas, by watermelon seed spitting contests.)

moded commissary) holding in his arms, Brack and Tillie's fourth (and last) child, and his two-year-old grandson, Ernie. To his right, almost obscured by men, is the mother, and to his left, the father—appropriately portly as the building's proprietor, though only in his thirty-ninth year. Presence of a second woman is modestly signified by the side view of a sunbonnet.[5]

On the gallery, Will Jackson, just arrived, looms tall, his neat straw boater conspicuous above a swarm of battered and rolled western hats.[6] His wife, Carrie Viola, would join him with their infant daughter, Onie Wilhoite, so soon as her father (himself a Jackson, she and Will being umpteenth cousins) considered her strong enough after having the baby to make the long train trip with him as escort.[7]

Northward, across The Field at Rose Cottage, Sallie Giles was not feeling well. The goiter on her neck was troubling her more. Dr. W__ in Austin called it a "tumor" and was prescribing an ointment, with morphine for the pain. She feared, horrible phrase, it might be a "pear cancer."[8] In her anxiety, she missed the closeness of her sisters: Amanda, who had lost her seventeen-year-old son, James, the year before and with Andrew about to be married would soon be alone with the younger children at the edge of the Mountains; Jane, in Austin; and Julie, far off in Montana.

Even with the Catos gone, Nocona continued to be a Mecca for youthful kindred. Mollie Clark's home was always open to her cousins. Besides, Jennie was there and also Julia Harriet (sometimes called "Hallie"). Julia, about the time their brother Charles married Vallie Billingsly and moved into a house of his own, had fled their Aunt Jane's determination that she marry her persistent Walnut Creek suitor, T__ W__. Once away, she had been quick to seek refuge from Jane's caustic teasing in the arms of Robert Dale Davis whom her aunt would never forgive for his part in thwarting her choice of a foster son-in-law. Hadn't she always (Jane asked) said that the girl would fly over all the pretty flowers and light on a cow pile! However much she must eschew the cattle culture when in polite mood, Jane could speak its language if plagued enough.[9]

---

5. This picture from Onie Jackson Kruger's files.

6. Boater: Webster calls this a slang term for a man's straw hat.

7. Onie Wilhoite Jackson (Mrs. Herman Kruger).

8. "Pear cancer" is how the writer, as a child, heard Sallie's ailment described.

9. This controversy between aunt and niece is still referred to in family circles.

In October, Alice and Laura entered Professor Albert Schuh's conservatory of Music and Fine Arts in the Temporary Capitol Building where their Uncle Jim, who had changed his business to real estate and loans, had a first-floor office.[10] The professor taught sketching from nature and also the Italian technique for reproducing old masters—a technique in which the Giles sisters and their Aunt Jane, who soon joined their classes, became so proficient in enlarging or diminishing to scale that their walls would be covered with the oil paintings they modestly disclaimed as "only copy work."

Christmas Eve, 1893, found Alice and Laura together at Mollie's house after a night's stop-over at Corn Hill (before the train change at Taylor) with their father's sister, Mrs. Elizabeth Land. The holidays, according to Laura's diary (which makes no mention of the absconded Mr. M__), were a joyous round of parties until she left the day after New Years. Alice would remain until March.

On the weekend of January 28, 1894, Sallie wrote Alice that Dr. W__ thought her tumor was a little smaller and a little softer, but, if the pain kept on, it had better come out—"to take it out would not be any trouble," he said. "However," she added, "he has made several mistakes before." She "would go on awhile longer with the ointment and the drugs." If she should change her mind, "the Maxwells" (Jim and Jane) wanted her to stay with them while she was under treatment. In which case, Alice must come home because "you or Laura should stay (in Austin) with me." Eugene "wished she would come on home anyway." He was being pressed into service as milkmaid and there were five cows!

On Monday, the letter continued, she drove into town to see the doctor again and spent the night with Lou and Val Giles. Back home next day her roses had been particularly beautiful. " . . . I don't think I ever saw them so large and double before. The yard was all over like spring." Then, in the night, a howling norther stiffened everything with ice. Nothing was left alive. Even Laura's pot plants and tubs of fern inside the house were frozen. The same was true with the three Daughterty sisters: Zackie's wife, Belle; Lillie and Vollie, all dedicated gardeners. Nothing they had was saved. Today, Sunday, was warm again, the sun shining, and would Alice ask Mollie, Julia and Jennie for some pretty scraps for Flora and Maggie? They had worked up everything in the house; Maggie was piecing Flora a quilt, and Flora

---

10. The Maxwell office is listed in the *Austin City Directory,* Professor Schuh's address is on his card in the Giles Papers and also on an 1893 program of classes.

was making a "tidy."[11] Flora had a whole envelope full of letters (scribbles) to send Alice. There was no other news except a lot of new babies. Mrs. Hayward Barr had had a girl and lost it. Christmas, Andrew Payton had married Eph's wife Margaret's sister Kate. His sister Eula (who was living in town with her Aunt Jane) had finished all her preparatory work and was now qualified to enter the University. And so the letter closes.

Eugene, when he went to Sprinkle for the mail on the 14th of February, two days before his fourteenth birthday, found a valentine mailed in Austin, bearing, such was the economy of the times, a one-cent stamp. The pink and white, beribboned missive was dainty enough for a girl's boudoir, hand-made and hand-painted in oils with a sprig of peach blossom under which, so tiny as to be almost invisible, was the single word "Love." In the center of the yellowed page is this stanza:[12]

May happiness ever be thy lot
Wherever thou shalt be
And joy and pleasure light the spot
That may be home to thee.

Laura's diary shows that on San Jacinto Day (April 21st), E. R__ was her escort for the maiden voyage of the *Ben Hur,* pleasure boat on the lake above Austin newly created by damming the Colorado. It was then that her Uncle Val, always the cut-up in Dock's family, saw fit, without orders from the captain, to pull the whistle cord, just when evangelistic-minded, four-times-widowed Mrs. W__, known as "the *shouting* Mrs. W__," (who wrote poetry as well as preached) flung her arms in exuberant welcome round a visiting dignitary. The blast against their eardrums, jumping everybody almost out of their skins, jumped the startled man right out of the widow's paralyzing embrace![13]

Laura and the erstwhile cadet had resumed their friendship, but with no return to tender sentiment. Between them, as between her and other young men whose attentions brightened the summer, including "Dub," her Aunt Belle's brother, and Mr. Walter Sanderson, whose sister Ada was among her and Alice's close friends, was always the shadow cast by that other affair, even when her aunts and uncles

11. Tidy or antimacassar: cover to protect a chair, where one's head rests, particularly from Maccassar or other prepared oils.

12. The valentine, with its ribbons, is among the papers.

13. The writer heard the episode reminisced in her childhood, but does not recall the raconteur.

ALICE

SALLIE

(in their mother's absence) took charge of Flora and Maggie at picnics and barbecues, seeing to it that they had the attention they required quite apart from the unmarried crowd.[14]

Laura was in her twenty-sixth year and Alice her twenty-eighth, but Black Land daughters with a prospect of goodly acres to enhance their worth were not being encouraged to marry early, though some did. In consequence, Tillie Birdwell's match at fifteen with her older sister Minnie's husband, Hayward's brother Brack, had never ceased to be a minor scandal among circumspect Texas and transplanted Virginia mothers, the latter permitting no formal dates until their girls were eighteen.

In September, Eugene and his twin cousin John tried for high school entry in Austin, but the one-teacher school at Gregg's Corner had not sufficiently prepared them.[15] They fell back to refresher work in Baker grade school. Quite willingly their Aunt Jane accepted them as roomers and boarders. For want of anything else to do in her spare time, other than visiting and painting, and perhaps because she was lonely, she had begun to take in young men students from the University which had moved from the Temporary Capitol into its own building across the street from the large house she and Jim had bought at 2009 University Avenue.[16] She needed no money, but she could never be idle and she liked to watch money accumulate.

When fall brought reasonably cool weather, Sallie felt she could no longer put off a drastic effort to relieve her suffering. On Friday, October 5th, Alice drove her into town, leaving Laura to cope with the problems at home and to walk the floor between tasks till she could hear that the operation had been successful. It helped to have Julia Alford Davis come three days later from Nocona, requiring that old Kate be hitched to the buggy for transport from one relative's house or old friend's house to the next. Thus, according to her diary, the time passed more easily till Sallie and Alice's return on the 15th, when Julia went into town to round out her stay with visits to her Aunt Jane and her brother Charles and his wife.

Soon Sallie was feeling so well that Dock took her to a Dessau

---

14. Jody felt the older girls had suffered from Sallie's determination that all her children should share and share alike.

15. Ledger No. 7, 1933-1935, p. 155; also Laura's diary.

16. The Maxwells sold this home in 1903 for $5,000. When the writer's father protested that the price was too small, his brother-in-law said it did not matter as the place had been sold in a good cause—the site for the Texas Bible Chair (which now includes the University Christian Church).

barbecue. On Wednesday, the 28th of November, she drove into town alone to bring Eugene and John home for Thanksgiving. Alice and Laura were preparing the dinner for both families and had invited as honor guest Professor Isaac Stephens, who was teaching the Dessau school. They would set the table for twelve.[17]

Approach of Christmas found Laura decorating a cedar tree, Alice preparing a costume in which to play Santa Claus, Dock looking for his copy of *The Night Before Christmas* which he would read aloud on Christmas Eve, to Flora and Maggie's delight. On the morning of the 24th, Eugene went into town for his cousin, Val C., Jr., three months younger than he. Christmas Day, after the excitement of opening their presents, the two boys walked across The Field for noon dinner with John and his brother Billie.

To calculate how often the turnrows between these homes were traversed would require programing on a computer. If the families were not dining together, or Sallie and Jody calling on one another, or the boys spending the night together, then Alice or Laura was apt to be week-ending with "Aunt Josie" (Alice's erudition having long since caused the younger set to drop the more familiar "Jody," though their elders never would, and to drop "Uncle Bud" in favor of "Uncle John").

In May of '85, Sallie had invested $60 in a parlor organ and in August of 1889 a like sum for a "Comfort Range" to relieve the kitchen drudgery.[18] The range her elder daughters appreciated for they were expert cooks, but music was not anything they craved. They had discovered their true metier when they took up painting.

Meanwhile, the summer passed with Alice presiding at the sewing machine which had preceded all other innovations in the house, Laura tending her mother's flower garden and chicken yard, the two sharing the preserving and jelly making from Dock's experimental orchard that bore peaches and plums in abundance, pears and apricots in profusion. They also shared the dairy work after Alice did the milking. Always waiting for any spare moments were the treasured easels—even Sallie, though she was beginning to feel ill again, occasionally dabbled with their paints and brushes; not, however, beyond a wreath of roses on a tin lid or a cluster of daisies on a felt dresser scarf. Which is failing to mention a continuous round of social life wherefor the house must always be in order.

---

17. Isaac Stephens, Sr. and Isaac Stephens, Jr. were both teachers; the former, who was William Gordon Bachman's son-in-law, taught for many years at Round Rock in Williamson County.

Following the 1896 New Year celebration with Aunt Jane and Uncle Jim and a valentine dinner at the Giles' in lieu of the twin cousins' customary birthday party, Laura must have found either time or paper scarce for there is a break in her diary notations. Or it may be she spent what time there was in reading—a list of books for 1895 includes novels from the classics, together with *Uncle Tom's Cabin* and a biography of Robert E. Lee. Then, too, she was under increasing emotional strain. Alice's visits to their Aunt Lizzie at Corn Hill were paying off in a manner not entirely to her sister's liking. Lee Dunn, no boy cadet or coronet player but a man of standing in the community—a teacher and something of a writer—was more and more the subject of Alice's conversation. Added to this, Sallie, who had seemed so well for awhile, was in renewed pain. The tumor was coming back.

Her frightened daughters persuaded her to go into town with them to have her picture made. They hid the cause of their fear by draping her in a shawl like some of the madonnas among Professor Schuh's prints.

Alice, extraordinarily skilled with her needle, watched her mother with an anxious eye as she whipped together row on row of insertion or stitched a thousand tiny tucks on the precious sewing machine. Not all the stitching she did was for herself or the family. To buy what she wanted, and her taste was exquisite, she sewed occasionally for others as she had since first getting the machine.[19]

As the year progressed, Sallie kept more and more to her bed. When Alice's betrothal to Lee Dunn was made public, Laura alone failed to rejoice. A future without Alice in the house was dreadful to contemplate. One might hope, but one could see that Sallie was sinking fast. She was, when the next year came in, reduced to writing household directions in a notebook she kept beside her, beginning them "My darlings . . ." A sentence addressed to no one, on an otherwise blank page, bespeaks her private despair: "I can neither walk nor talk."

---

18. Sallie's orders for organ and range are in the Giles Papers—the organ, Style No. 30001, bought from M. Kreisle Company, Austin, Texas, May 22, 1885, $70; the Home Comfort Range No. 9555 from the Wrought Iron Range Company, St. Louis, Missouri, August 14, 1889, $65.

19. Reminisced by neighbors. Much of her beautiful handiwork remains with the writer; some items, including a Persian lawn shirtwaist with row on row of insertion and 163 pin tucks, were claimed by a college junior granddaughter, Doris Wise, who declared the flounced skirts and even the hourglass corset perfect for the season's (summer of 1977) beach wear.

On March 28th, 1897, she died.

It was easier than it had been to get word to Uncle Val and Aunt Lou and Aunt Marguerite Henry in Austin, to Aunt Lizzie at Corn Hill, and to other distant kin, for Brack Barr had installed a telephone at the store which had become even more the community's messenger of death and disaster than a purveyor of his own business interests.

They buried Sallie in the plantation cemetery, to the right of the unpretentious granite pillar bearing Frederic and Harriet's names and dates.

Once the funereal stream of visitors became a trickle, Alice sought solace in her sewing whenever she could take her mind off Flora, who did not comprehend her mother's absence or the changes inevitable to a less patiently indulgent discipline. Emotionally overwrought, she was bent on the impossibility of matching her talented sisters' abilities and escaping their authority. Seeing Laura with a book would cause her to pick up a book on her own account and roar through the house till the rafters reechoed, "I'm gonna read! I'm gonna read! I'm gonna read!" Modern techniques might have taught her something, but such techniques were beyond her time.

Once she escaped the house and fled down the road, stripping off her clothes as she went. A kindly colored man, recognizing the situation, coaxed her into his buggy and drove her home. After that they locked her in a room, which did little to quiet the tumult. Although Alice's marriage would be delayed, Laura's reaction to its seeming inevitability intensified the gloom in Rose Cottage. She wept at any mention of trousseau or wedding. She had nothing against Lee Dunn, but to see Alice occupied with bridal ruffles and flutings brought more than tears. It brought bitter rebellion at thought of being alone with the shadow.[20]

The summer passed somehow. Autumn was helped by organization of a Tuesday Book Club over on Gilliland with Laura and Alice added to the roll of twenty Manor and Gregg young ladies at its second meeting.[21] What more welcome anodyne for hurt and sorrow? The first norther came. The traditional shoat was killed. A fruitcake was baked and shut in the pantry to wait for Christmas. By and by a cedar was cut and brought in to be decked with candles and candies.

For Flora and Maggie, Santa's visit was but hours off when Mr.

---

20. Reminisced by relatives.

21. Imogene Kuempel Acklin (Mrs. J. E. Acklin, Round Rock) has the scrapbook record of the still extant Tuesday Reading Club.

LAURA

Barr sent a rider to tell Alice she was wanted on the telephone for an urgent message from Corn Hill.

None present at the store, including four-year-old Onie, would forget the moment of tragedy when Alice learned that her betrothed was dead. Not only was Lee Dunn gone, stricken by that swift killer, typhoid fever, but his brother, Will, also. The bereft parents, looking on her as a daughter, needed her. How soon would she come?

Days later, she returned to her father's house, clad in mourning. Not widowed in name, she could at least be widowed in spirit. The trousseau was banished to the attic where she and Mollie and Laura had climbed to a makeshift girls' haven before Dock added two rooms below. Yards of black replaced the bridal stuff in her room.

Over and over she read the copy she had made of the article Lee Dunn left unfinished:

> That our soil is peculiarly adapted to the production of cotton is too obvious to demand even a passing notice here, for it is shown in the conclusive fact that we are its first competitors in the markets of the world. But it is the first factor in our destiny and development . . . It is at once evident that the development of a country depends not only on the class of its people, but also upon its natural and acquired resources.[22]

She mourned his loss, both for herself and for all the little communities round about that had lost a man who was a thinker. The gold band that was her betrothal ring must never leave her finger.[23]

Sprinkle was the up-and-coming community of the moment; Fiskville had been supplanted as Walnut Creek's most popular postal-and-shopping center. Zimmermann's store, burned down, would not be rebuilt.[24] It was Manor, over on Gilliland, that had to be reckoned with. Manor, that had been just "Wheeler's Store," had mushroomed with the coming of a railroad. Now it could boast several business establishments, half a thousand families, and seven churches, two of them colored, one Swedish Lutheran.[25] The Swedish migration, brought on by legislation opposed to patriating Blacks, was sweeping that side of the county.[26] Never mind, Blacks, Swedes and old planter families together were producing enough cotton to support the Manor gin that had replaced little gins burned along the two branches of the

22. Alice's copy is among the Giles Papers.

23. Reminisced by relatives.

24. Ledger No. 9, 1930.

25. *Texas State Gazette,* 1896-1897.

26. Fannie Crockett, "Parson's Female Academy" (at Manor) ms., Austin and Travis County Collection.

Cameron Road, all except the Dessau and Sprinkle gins.[27] What Sprinkle needed now was a railroad of its own.

Men lounging about the Barr store Saturday afternoons speculated, between domino games, horseshoe pitching and spitting contests, on the possibility of getting a railroad, but talk grew heated only over the price of cotton in the bale or over how many acres the coming season would see dedicated to the reign of the anointed king.[28] No one thought of the place, though Lee Dunn might have seen it so, as a Virginia outpost on Texas' cultural frontier.

---

27. Eugene's Memoirs mention a number of such gins besides the Maxwell gin: the C. C. Baker gin at Gregg's Corner, the Aaron Boyce gin, the Edwards gin on the "old Cain place," the Dederich gin at Dessau, most of which burned.

28. Status of tobacco spitting as a social art in Virginia culture merits repetition of a story told the writer by Miss Anna Simonds, daughter of Dr. Frederick Simonds, long the chairman of the University of Texas Geology Department. The Simonds were Episcopalians and neighbors of Bishop George Herbert Kinsolving. They were bringing up their small grandson and nephew, who, one mild winter morning, made inquiry concerning the bishop's ring. His aunt suggested that he go visit the bishop for an hour and get the information first hand.

In an hour he was back, filled with enthusiasm, but not for the ring. "Oh, Auntie," he burst out, "you just ought to see him spit!" His Virginia Excellency had entertained with an exhibition of such skill in aiming choice *ambre* from various distances to sputter in points of his fireplace flame as to drive all thought of the visit's original purpose from the child's mind.

THE WILL JACKSON family before moving into the Gingerbread Mansion—(l to r) Elmo with dog, Onie, Ola, Fred, Will.

Gilesburg: The Novice Entrepreneur

"Sheene, Sheene, have some scheese!
She stink a liddle, but she's goot!"

—Customer from the Old Country treating to limberger cheese and soda crackers in the new store.

From an Indian point of view, the Black Land was being scalped as the grass disappeared, leaving Mother Earth ready for a cotton wig.[1] But the Indian was gone and the Europoidal settler who succeeded the wandering tribesman boasted an ability to plow a straight furrow a mile long. The ripped earth, biding its revenge, offered an abundance of foodstuff, but no outlet for youthful ambition other than the plow except to "go west" like Ocie Cate—or go into politics. Unless, of course, like Brack Barr, one set up a mercantile establishment on one's own holding.

Across the Mississippi men of the gentry, broken by their own myth of military might, had come home from the war to idle on their ruined plantations or work with their hands or watch their women do it; but that other kind of Southerner, the Red Neck,[2] snatching the opportunity afforded by the northern idea of democracy, set about an instant politics based on Louis XIV's "one drop of blood" theorem,

---

1. To an Indian, grass was Mother Earth's hair—Mother Earth, whence he sprang, sustained him in the light of the Sun, his father, and took him—at least his body—to her bosom when he died.

2. Red Neck: expelled Scots and British, working in the tropic Caribbean area in their kilts, blistering in the sun, were derided as "Red Legs." When they moved onto the mainland and were later protected by coveralls, the term became "Red Necks," synonymous with "white trash." Today, in the South and Southwest, calling a person "trash" or referring to anyone as "poor white trash" is an insult, deadly as an obscenity.

already deep rooted in provincial laws of Louisiana—an area where that amoral and ignorant but ambitious monarch had thought to build a cotton power structure just when the Christian conscience was about to disavow the slave system.[3]

Clutching a Bible in one hand, a flaming torch in the other, the Red Neck demagogue swept the South, proclaiming a "lily-white" doctrine, boasting his own poverty and sun-blistered skin, shouting with evangelistic fervor, "Ah'm ah pore man, ah am! Ain't got nothin', Ain't never had nothin'! But ah'ma *white* man, by God! Gimme this office an' won't no damn nigger git it, you bet! Nor no damn Black Republican, neither! You wan'a be run over by niggers and Republicans?"

The Red Neck never once mentioned Great Louis, who had provided him his thunder. Probably he had never heard that much history. Nor could he read the Bible he carried, except to parrot preachers of his own ilk who gave its words a literal meaning to justify the superiority of poverty (inclusive of ignorance) and of a skin minus the pigmentation to prevent its blistering in the sun. Thus the woefully fragmented South was due to be overcome a second time, this time by a social disease it failed to recognize or was too down-in-the-mouth to care about—a disease called "racism."[4]

As for Texas, Eugene comments in his Memoirs:

> Through father and other Jeffersonian Democrats, I was thoroughly inoculated with the principles of southern democracy they held to and fought for—how far, far afield we later have gone from many of the fundamental principles they so conscientiously and ardently advocated.[5]

For mature men like Dock, the Democratic Party was a vehicle to restore states' rights lost in the ill-fated venture of the Confederacy.

---

3. Louis XIV, one of Europe's most unlearned, aggressive and egotistical monarchs, full of archane superstitions, derived the formula that all black things are of the Devil and that man's physical heredity lay in the bloodstream—therefore even one drop of his Satanic Majesty's blood must reduce a lineal descendant to the level of the beast with no hope of attaining a higher estate in the order of creation. Hence the absence of a soul in any African descendant, no matter how far removed.

This formula is nowhere stated in these exact terms. It is derivable from a phychological analysis of the biographies of the Great Louis. See Atkinson: *Creolism and Racism on the Gulf Coast,* 1977.

4. Racism: ethnic hostility; derived from the Latin *radices* (roots), therefore relating to differing heredities, both physical and cultural.

5. Ledger No. 12, 1942-1943.

For others, like Zackie, too young to have worn the gray but old enough to remember brothers and cousins in uniform and the Iron Clad Oath that had disenfranchised those who survived, it was to become both a guidon and a shibboleth, a social machinery designed to save the country and them personally from the down-hill pull of post-bellum life.[6]

Dock was Sprinkle's election judge in 1892—the year of the Hogg-Clark gubernatorial controversy—and Eugene, aged twelve, spent election day at the polling place with his father. Ballots were cast in the old commissary because Brack Barr had not yet expanded the business into a general merchandise store.

Verbal warfare had been hot. Among battle cries the Memoirs list are, from the judge's side: HOGG, HARRISON AND HELL! CLARK, CLEVELAND AND CHRIST! and from the governor's side: CLARK, CUNY AND COONS![7]

In Eugene's mind, wrought on by his elders' emotionalism, veterans were more sanctified through blood spilled in the Lost Cause than were saints exalted by evangels coming shouting out of Dixie to close the cotton season with exhortations to repentance and promise of salvation as baskets were passed to collect cotton pickers' money. Surely, he thought, heaven must fall—even as the Old South—vanished like the Holy Grail—if Clark (heading the Populist Party that had split off from the Democrats) failed to succeed Hogg in the governor's chair! Accentuating his fears was knowledge that in Austin his Uncle Jim Maxwell would, like others he knew and loved, be voting for Hogg.[8]

The day was just dawning, overcast and chilly, when he and Dock drove across The Field with a box for the ballots. To further dampen his spirit, a cold drizzle started toward noon. Clark people built

---

6. Iron Clad Oath: see Chapter Five, Note 6. Guidon: a flag to mark the way. Shibboleth: a tribal title and distinction, biblical.

7. Cuney: Norris Wright Cuney, born 1842; son of a Hempstead, Texas, planter, Col. Philip Cuney, and a slave mother, Adeline Stewart. Freed by the master, Adeline sent her son to a small school for free colored boys in Pittsburgh, Pennsylvania. In 1896 he was collector of customs at Galveston, appointed by President Harrison, and the political leader of the state: biographical note, courtesy of Dr. J. Reuben Sheeler, head of the Department of History, Texas Southern University, where the writer was an associate professor for ten years. Coon: presumably from "barracoon" where "black ivory"—captive Africans—were confined waiting shipment overseas.

8. Hogg-Clark Campaign: Ledger No. 8, 1936-1937.

9. See voting talley appended to these footnotes.

themselves a fire under an open lean-to. As the weather worsened, Hogg people, decidedly in the minority, sheathed their verbal stilettos and, holding anathemas in abeyance, came, diffidently, to share the opposition's meager warmth.

Late in the day the boy discovered his Uncle Jim driving out from town in spite of the weather to see how the Sprinkle vote was going. Also, old Buck Tate (colored) was ambling up, shoulders hunched against the rain. The buggy halted. After a brief conversation, the old man nodded assent, then he and the buggy moved forward. But as his Uncle Jim sought a hitching post, Eugene saw his Uncle Zack rush out to intercept the prospective voter. Again a brief conversation and an assenting nod, after which Tate went in and cast his ballot for Clark as he intended to do all along.[9] Nevertheless, though Sprinkle went overwhelmingly for Clark, thanks to Black support, when the full vote was counted, Hogg had won statewide—and heaven did not fall!

Six years later, Zackie had reached the middle forties, a prosperous farmer and rancher through inheritance and wise investment. He had been elected both county commissioner and justice of the peace. With a head for figures like his cousin Tobbie—a pet name for Joseph Andrew Bachman, William Gordon's son, who was an inventor holding patents on machinery—he, like Joe, wanted something out of the ordinary.[10] He and Joe and Ocie Cato were men of The Interlude—men between Eugene's oncoming generation and Dock's generation of veterans who were fast becoming mystique and legend. He knew he had a talent for vote-getting and his eye was on the county treasurer's seat.

It would be some years yet before institution of the White Primary would end all hope for the Republican Party in Texas by emasculating the Freedman as a voter.[11] So far as that vote was

---

10. The writer once asked her elderly cousin, a man of distinction and wealth (who must have been a dreamer as a boy), why he was known in the family as "Tobbie" or "Tob"—the answer: They said when I was growing up if I were ever to amount to anything it would have *to be* in the future "because it sure wasn't now!"

11. Passage of the Terrill Election Laws, effective in 1905, limited the Democratic Party's nominative straw vote, called the "Primary," to "white males." With Democrats, even minus the Black vote, heavily in the majority, this created a Catch 22 situation on a purely complexional basis and forced a colored man to choose, in the general election at the year's end, between the Democrats' choices for local and state office, who were certain to win, and the Republican nominees who were equally certain to lose. As a consequence, the last Black legislator descended the capital steps in 1905 until the so-called

concerned, Zackie knew the post-bellum Black to be the same fellow he and his brothers and sisters had grown up with—a person totally unlike the South's Red Neck who had grown up hating both master and slave because both looked down on "white trash"; he knew how Texas Blacks had stood by to protect women and children and operate the plantations till the war was over and how, afterward, bonds of intimacy had not been broken. But politics is a pragmatic game. And there was that tidal wave from the bankrupt South flooding into Texas where unbroken sod could still be had for a few cents an acre—a tide bringing on its crest Red Neck slogans. Already the Jim Crow idea was launched and though Zackie fought shy of it through the Hogg-Clark rupture, he could and would adapt Red Neck rhetoric to his personal need wherever possible while continuing to woo voters of all complexions.[12]

Down Webberville way, campaigning with his good friend, George Calhoun, whose objective was a judgeship, he began a "pore-mouth" peroration with: "Ah'mah *pore* man; ah've worked hard *all* of mah life . . ." breaking off to point a finger at a Black listener standing against the canvas of the campaign tent, ". . . and there," the resonant voice picked up after a second of silence as necks craned to see who was at the back, "is the man who can tell you so, for he's known me *all of my life*!"

"White Primary" was abolished in 1940, thanks to the efforts of Houston's colored dentist, Dr. Lonnie Smith.

12. Jim Crow: originally a comic character in an Atlanta minstrel show—a small Black boy carried in a sack on the master's back—the refrain of the act: "Wheel about an turn about an do jes' so an' evah time you turn about, it's 'Jump, Jim Crow!' " Reflecting the deep sense of class in the South, Mrs. Leonia L. Dorsey, a retired librarian and English teacher, proof reading *Kingdom,* paused to say her father, Wm. Henry Lanier, son of the master of Lanier Plantation, pictured for her, from his childhood in Alabama, Blacks too tiny to be allowed underfoot in the way of field workers; dressed in flour sacks with holes cut for sleeves, they might be seen ranged on rail fences and heard chanting to themselves:

> Mah name's Sam*
> An' ah' don' give a damn!
> Ah' *druther* be a niggah
> Dan a *po'* white man!

*Sam: from Spanish *sambo*—humorous designation of fair-skinned slaves like Lanier. Sambo: from *zhambo*—Kongo word for "monkey"—used in America to denote Indian-afro genes.

Later, when the crowd poured out into darkness, the Black man (was it Wiley Moore?) plucked at the Calhoun sleeve to say softly, "Ah shore was glad Mistah Jourdan didn't ast *me* to say nothin', 'cause effen he had, ah'd ah had to say if walkin' across a field wid a gun ovah his shouldah an' a little dog follerin' at his heels was ha'd work, he sho' done plenty of it!"[13]

Eugene would be deeply interested in his uncle's campaign by fall of the year 1898, but February and his eighteenth birthday saw him still more interested in the thought that Brack Barr, at age eighteen, had founded a thriving mercantile business. True, the Barr store a couple of miles southeastward and the Gregg store an equal distance northeastward were both doing well while the store at Dessau directly down the road from his own front gate had failed and closed up; nevertheless, right behind the house lay The Field, cut into farms occupied by kinnery and familiar Blacks, the latter either third-and-fourth renters owning their teams and tools or owners of the acres they tilled—a prosperous community in itself. Why not a try at a new center on the Dessau spur of the Cameron Road?[14]

He confided his aspirations to his father, who, while commencing to live in dignified retrospect, was yet too sensitive and too wise to wet blanket his son's ambition. Not only did Dock agree to sponsor the project, he encouraged it. Among other reasons, it would be something to take Alice's mind off her grief and help the rest of the family surmount the gloom of her "widow's weeds." In sympathy with her feeling, he was not in sympathy with so dramatic an expression of it.[15]

Manor, nearer the center of the county's farming industry than Austin and beginning to rival Austin in population and business, was also nearer home. Tools, nails and hardware were contracted for with the Manor Lumber Company to build a gable-roofed store sixteen by thirty with 480 square feet of floor space. Neighbors coming to help promised quick construction but the roof was scarcely on when unforeseen trials and tribulations set in. Eugene went down under onslaught of the killer fever, typhoid, and was saved from death through the combined effort of the elderly Dr. Shannon from Fiskville and the new Dr. Fields from Manor. Maggie backed too close to the open fire in the dining room. Her clothes flamed. Alice, beating the blaze with her hands, caught fire also. Laura, smothering the conflagration with

---

13. Reminiscence: Judge George Calhoun, Austin, to writer.

14. Ledger No. 6, 1930-1932, gives the story of Eugene's first mercantile venture.

15. Miss Annie B. Giles, reminiscing to writer, 1976.

HOGG—CLARK GUBERNATORIAL RACE, 1892
Sprinkle Box
(from W. L. Giles' notebook)

Blacks: (for Clark)	Non-Blacks (for Clark)	Non-Blacks (for Hogg)
Richard Bradshaw	James Dederich	M. Yager
Joe Davis	James Dederich, Jr.	Wm. Boyce
Jem Clark	Ed Leverett	Aaron Boyce
Will Shanks	Sugar Cain	Ike Evans
Tom Madison	W. L. Giles	Will Howell
Bill Black	J. B. Jourdan	?J.M.P. Kirland
Buck Tate	John Almquest	Haywood Barr
Green Edwards	Eph Payton	W. B. Barr
James Peoples	Z.P. Jourdan	E. F. Sprinkle
Tom Bratton	Sam Greene	John Rowzee
Wiley Moore	Dan Penick	Arthur Jarmon
Commodore Jesse	Steph Jarmon	Bill Perry
Will Clark	Walter Jarmon	W. J. Harrison
Dow Clark	Ed Jarmon	John Hines
Newel Binford	T. J. Rowzee	M. B. Kino
Bill Binford	Oscar Stephens	E. P. Penick
Ben Baker	Walter Sanderson	Tom Smith
Will Baker	J. H. Sanderson	Charles Lawson
Cal Jackson	C. L. Morris	Ches Huffleur
Frank Jackson	?Song Johnson	T. J. Weaver
Frank Robinson	S. B. Giles)	John Foster
Salem Robinson	W. B. Giles) not related	Dub Daugherty
Silas Simms	J. C. Giles) to	John Daugherty
Robert Hill	B. C. Giles) W.L. Giles	Joe Clayton
Tom Dickerson	James Foster	James Clayton
Prince Foster	Claud Johnson	Old Brown
Button Crayton	____ Murphy	John Blach
Jim Davidson	____ Polegshot?	____ Boaman (Bowman)
William Washington	____ Sneed	____ August
Husing Clark (Housen Clark)	____ Sneed	____ Seiders
Columbus Bradshaw	H. A. Dorris	____ Siders
James Binford	____ Groscup (Grosskoph)	____ Seiders
William Jones	____ Schroder	____ Mill
George Plummer	____ Schroeder	____ Helger
Nathan Vance	____ Groman	Bob Kino
Charles		
____ Jones		
____ Bradley		

These listings show how some families were divided. Question marks show names where reading is uncertain.

GILESBURG Store—on gallery (l to r) Kie Boyce, Billie Jourdan, John Jourdan, and Dock Giles.

quilts, was painfully burned. All that would heal, but Flora was growing more and more unmanageable. Not knowing what else to do, they set her apart in a little house in the yard—Frederic's commissary that Sallie had moved up from the Old Place; there each day they took her what they thought she might use and appreciate without becoming violently destructive: food (no knife or fork), bright scraps of fabric to finger, bouquets Maggie gathered for her from their mother's flowerbeds or from the pastures. They took her for walks when the family was alone but never again in her almost sixty years would she be seen by the public eye.

At Christmas, Dock could stand Alice's mourning garb no longer. He demanded in the name of common sense that she take it off. He said that, sad as Lee Dunn's death was, the world had not come to an end for her because of it. She laid the black aside, but not her bethrothal ring. Nor would the ruffled bridal dress ever come down from the trunk in the attic.

It was the 15th of June, 1889, before a sign went up on the store gable:

GILESBURG
U. S. Post Office

---

16. Z. P. Jourdan held this office from 1899 to 1906, when county finances were changed to a different system and the treasurer's office abolished.

Advent of his store, as the novice entrepreneur puts it in his Memoirs, "with me as postmaster and father as my assistant," his Uncle Zack's election as county treasurer and his Uncle Ocie's as sheriff of Custer County, Montana, were not the only family sensations of the closing decade.[16] It had been assumed that Leo's name in 1892 brought the Bible birth page to an end, but a year after Sallie's death, John and Jody had astonished the countryside with a daughter—the first "Miss" Jourdan since George and Lucy's daughter Frances in 1861. The child was now over a year old.

14

The Big Blow

The chief characteristic of weather in Texas is not variety but surprise.

—Sir Swante Palm

A Texas norther is to a Gulf hurricane is as a ten-year-old bully to Muhammed Ali, who used to be Cassius Clay.

—M. Atkinson

A cotton year was November to November.[1] November of 1899 came in wet, putting a promising season in the ground. The rains continued through the winter and spring; planting had to be done in mud. And still it poured down. Late planting made no difference in the crop yield—the unprecedented moisture brought the cotton up in a hurry. And the weeds with it! The problem was to get in choppers fast enough.

Cultivation between planting and picking was in three phases: thinning, chopping and "running out the middles." Planted thick to ensure a "stand," baby cotton was reduced by field hands swinging hoes lightly to cut out every other hoe-blade length from the green ribbons of two-leafed cotyledons. This was "thinning." In a few weeks they went down the rows again, taking out extra cotton, grass or weeds. This was "chopping." At the same time, or a little later, a plowman, perhaps on a "riding cultivator," such was progress in the cotton industry, cleaned the furrows between the rows. This was "running out the middles."

In the dividing and subdividing that had made The Field into

1. In recognition of this fact, Texas law forbade changing tenants on the part of landlord or other entrepreneur after the middle of November—this because of recognized abuses by some who had driven tenants away just in time to themselves gather the crop.

heirships and so into individual farms, Alford land lay next to Giles land. Charles, with his thriving grocery business (the only one of Mary Catherine's children to be established in Travis County), accommodated to the situation by buying out the interests of his siblings.[2] It had been simple enough with the girls, who were together in Nocona; not too difficult with Baldy, who had married a New Orleans girl, Katie Smith, and settled permanently in her town; but Tom, the eldest, cow-boying with his Uncle Ocie on the Yellowstone in Montana, was at times difficult to locate for a necessary signature.[3] Nevertheless, by New Year's of 1900, the Alford place was legally well in hand and rented to a farmer, Seth Thomas, who was also a blacksmith. The Gileses built Thomas a shop next to their store and, looking forward to a more than usual influx of transient cotton choppers, come spring, added a room to the store itself for housing bulk staples.

So crying was the need "to get the cotton out of the grass" when spring came that, like fields gone wild in the rain, the store was overwhelmed by strange customers, not to speak of neighbors coming in to wait while a horse or mule was shod or a plow point sharpened. Eugene, faced with shelves empty except for a single jar of snuff and a few peppermint sticks and curls of ribbon candy, scraping the bottoms of barrels in the storeroom, saw it was up to him to get replacements, mud or no mud.

He now had a telephone over which Austin wholesalers assured him his credit was good if he could make it into town. The Cameron Road was a loblolly. To overcome hub-deep mire, he hitched four mules to a light spring wagon and started, leaving his father in charge of the mail and the counter. His sisters would keep the duebills totted

---

2. The writer has been unable to find any deed of distribution or conveyance among the Alford heirs concerning an acreage in the Mays Survey Frederic Jourdan deeded to his daughter, Mary Catherine (under the name of M. C. Sprinkle) November 16, 1880, Deed Book 48, pp. 94-95. This tract lies along the south bank of Walnut Creek, commonly called Polecat Creek, which bounds one side of Lot 20 in the Harriet Jourdan Survey, see map, Appendix D. This gift was made less than a month after Mary Catherine Jourdan Alford became Mary Catherine Sprinkle. She died four and a half months after the deed was signed. Since her children were Alfords, her estate was carried in the name of Alford. Presumably her heirs never recognized the existence of this particular deed because of the name difference, although Charles continued to pay taxes on the acreage until 1928.

3. Deed Book 100, p. 635; April 7, 1891, T. F. Alford to C. F. Alford, all interest in Frederic and Harriet's estate for $350 cash, Yellowstone, Montana.

up.[4] As it continued to rain, this family help allowed him time both for necessary trips to town and for the cultivator, uprooting green stuff between the rows lest it choke the branching cotton plants. He could look back and see water running down the furrows behind him, unaware of the greater threat to the earth loosed with every step the mules took.

Over at Sprinkle, Brack Barr, watching the promise of the fields, tore out his old gin and more than doubled its capacity by building what the Memoirs describe as "two gins under one roof or two batteries of five gin stands each with separate presses."[5] At the same time, he enlarged both his store and his blacksmith shop, set up houses for his employees and began what was, for so financially conservative an area, the stupendous extravaganza of a gingerbread mansion with a corresponding spread of outbuildings trimmed in "sawmill lace."

Down the country at La Grange, seat of Fayette County on the Black Prairie, Henry Thomas, a gentleman farmer, even more cautious with his dollars, perhaps, than most Black Landers, broke his hitherto unvarying rule: to scramble a metaphor, he banked on his crop before it hatched. He ordered, paying $500 cash, a piano for his only daughter, Mary, who was of an age and temperament to begin music lessons.[6]

The instrument had not yet arrived, carefully packed, when the Austin dam broke (April 7, 1900) and the Colorado River overflowed, washing everybody's expectations into the Gulf of Mexico. The only hope left was that another planting, late and expensive as it would be, might mature before frost fell.

Black Landers' cotton, better drained and above the flood, went right on growing in the rain till some said it was tall enough to hide a man on horseback. Even so, joy was not unconfined. Though stalks bent under weight of hard, pointed green bolls that normally would augur a bumper crop, the foliage was so overlapped and entangled that it cut off sunlight needed to mature the fibres inside the green casements. Fear was that mildew and mold might take over before the protective hulls burst into dried brown husks, exposing a wealth of fleecy locks.

---

4. Duebill: a slip recording a purchase to be paid for "when the crop is in."

5. The Memoirs say both sides were almost never operated simultaneously.

6. Mrs. Mary Thomas Simons, Austin, reminiscence to the writer, 1977.

Still, however much anxiety might hold anticipation in check, cotton growers must hire "hands" and stand for their living expenses and local merchants must issue duebills for those expenses, running up accounts with the wholesalers who would have their own obligations to meet eventually. The gamble of cotton could never be other than a domino operation.

On September 8th, when the bolls seemed about to open, befell Texas' tragedy of all time—the great Galveston storm—a hurricane that came without warning to wreck the port, taking lives in countless thousands. As the mighty wind swept inland its velocity lessened but it was of sufficient force when it reached Fayette County and the Black Prairie to wipe that second stand of cotton off the earth, leaving the fields stark and bare.

It was different on the Black Land proper. After a night of buffeting by the Big Blow, people opened their doors on a seeming miracle: The wind, its strength diminishing with every inward and upward mile, had maintained just enough force to shear away foliage without disturbing the bolls.[7] Very shortly, a hot sun sizzling down brought the fields to a crispness like gargantuan frying pans filled with the whiteness of popping corn—the South's "summer snow."[8]

Out of respect to the horror on Galveston Island where, pressed by the city's terrible emergency, bodies were being loaded three deep, crosswise, lengthwise, then crosswise again on "cotton crawlers" (flatbed trucks for conveying cotton bales from train to boat) and hauled to funeral pyres, there was no jubilee—not even spoken reference to the ancient adage, "It's an ill wind that blows nobody any good."

---

7. Ledger No. 8, 1936-1937. Essay in old tablet, n.d.
8. "Ethnogenesis," Henry Timrod; South Carolina, February, 1861.

15

The Gingerbread Mansion

The bees flee hame wi' loads o' treasure
The minutes wing their way wi' pleasure.
—Robert Burns, "Tam O'Shanter"

The Memoirs say 1900 was the big cotton year of all time. The average yield was a bale to the acre, of high grade and fine staple. The only part of any year farmers were not buying on credit was in the fall when the crop was in but this was a fall to lay all other falls in the shade! Eugene thought one of the German pioneers put the situation in a nutshell: "I vent to Austin and everywhere I vent, in the stores, on the streets, the mens had bundles, the vemens had bundles, the childrens had bundles! Christ! I never for seen so many bundles!"[1]

The Gileses added a room between the store and the blacksmith shop to house bulk supplies.

Almost overnight, the Memoirs say, "prosperity was abroad in our land." It began with those semi-nomads, the pickers, who had "never had it so good." Competition for help was keen after the storm for time was of the essence—a return of the rain could dash the fleecy locks from the spread-open bolls in a second, and in the mud all value would be gone from the "vegetable wool." Price paid for picking shot up from "fifty cents a hundred (pounds) to a dollar."

Ordinarily, pickers crawled furrows between the rows, their knees protected by heavy leather pads, or bent their backs to pull the fleece from the crisp, dried husks. In this extraordinary year, they reached up instead of down. In any year, good pickers, if men, tallied three hundred and fifty to four hundred pounds at the end of a day; if women, two hundred and fifty pounds. Champions might go as high as seven hundred pounds and there was talk of "boss pickers" achiev-

1. From the Memoirs: Tablet Essay No. 1, p. 18. The next two quotations are from the same essay.

ing a thousand pounds.[2] "Weighing up" was regularly twice a day, before noon and at sundown; oftener for the best pickers. They came in ethnic procession to a steelyard (scale) suspended from an arm on the back of a wagon, each with a canvas sack a yard square, five to seven feet long (shorter for children), dragging from a single over-the-shoulder strap; and they remained ethnically queued, Blacks with Blacks, Latinos with Latinos; Poor Whites, if any, with Poor Whites—not because they had to, but because each preferred to be in familiar company.[3]

"Fortunately," the Memoirs continue, "the selling price was upped ten cents per pound," which means cotton brokers were paying ten cents a pound instead of eight for cotton in the bale—that is, ginned or "lint cotton."[4] Eighteen pounds of seed cotton averaged a five hundred pound bale of lint. At this rate, a family on a "small farm" (usually 50 acres) able to do its own picking, with an acreage for forage crops and the price of ginning subtracted, came out with $2,000 to $2,500 for their year of labor.

> Everyone (still quoting) except a few tightwads went on a spending jag. Honest people paid in full for what they were buying, if it were possible, but where incoming cash (after they settled their duebills) would not cover the price of a fringed-top surrey or a fancy parlor organ, the same dealers who were backing the supply of Gilesburg merchandise most graciously extended credit—at 10% interest.

To the men and women on whom the king's regime had imposed the swing of a hoe up and down greening rows, the drag of a sack up and down furrows, the pull of plow reins—and to children for whom Santa never before had side-tracked his sleigh—it was as if the millenium traveling evangels promised had arrived ahead of schedule.

---

2. The writer's memory of individual pickers has been reinforced by interviews with a number of persons in Austin and Houston who recall the days before the Rust brothers (one of whom, Mack D. Rust, was best man at her wedding) invented a mechanical picker. Mrs. Augusta Nelson of Taylor and Austin remembers "a man named Sorrell" as among those picking a thousand pounds a day.

3. Some of the most picturesque scenes in the winter's memory are these *genre* groupings.

4. Cotton has been known to sell as low as four cents a pound. The writer remembers one down period when her father's voice reverberated over The Field for what must have been a mile, maybe two, and not without profanity, that he would "roll my cotton into the Colorado River before Moody (the cotton buyer from Galveston) can have it at such a price!"

Temptation to buy because they could, and so raise themselves a step or two in the social scale, was irresistible. Women with houses too small for a parlor replaced the pieced quilt on the bed in the "front room" with a snowy Marseilles spread dropping almost to the floor.[5] They put aside their sunbonnets on Saturdays and Sundays in favor of hats loaded with feathers or flowers and their men drove them to town or church in top buggies instead of old spring wagons.

The next year (again quoting the Memoirs) nobody outside Austin rivaled the Barrs in splendor. The gingerbread mansion with outbuildings to match was furnished in keeping with its architecture. Downstairs, doors slid back into the wall so the rooms could be thrown together for parties and receptions. Tillie hired a staff of house servants and Brack a colored driver for their "very stylish, rubber-tired vehicle (presumably an underslung phaeton) drawn by matched black horses in silver-mounted harness."[6] Meals were sent out to eighty-eight-year-old Erasmus Sprinkle in his room (once the commissary) lest he track the polished floors.[7]

Older heads among the gentry eyed this burst of glory dubiously. Mamas, growing more and more socially conservative as land values rose in accord with cotton prices, bowed coolly to occupants of the handsome equipage rolling past. Nor did they hasten to leave calling cards on the silver tray in Tillie's drawing room except when Mother

---

5. Marseilles: a heavy cotton cloth first woven in Marseilles, France. Bedspreads of this fabric, patterned always in white with fringed borders, were exceedingly popular at the turn of the century.

Homes of small farmers and third-and-fourth renters in Central Texas were not so miserable as the homes of share-croppers described by C. T. Carpenter in "King Cotton's Slaves," *Scribner's Magazine,* October, 1955; nevertheless, they all too often lacked "any touch of beauty to relieve their dreary barrenness" unless it were a few pots of geraniums alongside a brass-bound cedar water bucket on a shelf outside the door and a tin molasses bucket dropping strands of curly moss, suspended, together with a saddle, from a rafter of the shedded front porch.

What is known as a share-cropper in the South—one to whom everything from daily maintenance to teams and tools was furnished—was called a half-renter or "halvie" in Central Texas, since crops were divided equally between renter and landlord who stood good for his tenant's duebills. Above the half-renter, quite frequently an entrepreneur of considerable status in a community, was the third-and-fourth renter who furnished everything himself, the landlord receiving but a fourth of the cotton and a third of the forage crops.

6. Ledger No. 9, 1937.

7. Reminiscences, Mrs. Onie Jackson Kruger, Austin, Texas.

THE FOUR GILES MEN—(l to r) Dock, Eugene, Val C. Jr., and Val C. Sr. with dog.

GILES DINNER PARTY at Rose Cottage—Alice, Val C. Jr., Minnie Greenhill Giles, Laura, Annie B. Giles, Dock, Val C. Sr.

Barr, a ponderous lady, popularly "Aunt Cree" or "Cousin Cree" to a host of young relatives, left her home in Austin with her daughter, Ida, to visit her sons and grandchildren. Ida was married to Oscar Stephens (no known relation to Professor Ike Stephens) who took his chances where he found them—but not in cotton.

Those who concurred in Sir Swante Palm's dictum on Texas weather said openly, "The Barrs are putting on airs!" and privately, "All that goes up must come down!" These were the folks who remembered that, in the midst of their own good fortune the year before, only sixty miles down-river, cotton on the Fayette Prairie had been completely washed out.

For those who know nothing of the cotton industry, it may be well to state that "grade" refers to the condition of lint cotton—that is, cotton that has been ginned to clear it of seed—as to whether it has been discolored by weather or insects, has been dirtied by being blown out of the boll on to the ground or carelessly picked so that bits of brown hull are mixed with the fibres. "Staple" refers to length of the fibre which may vary from three-fourths of an inch for very short cotton to an inch and a half for long staple.

David L. Cohn, in *The Life and Times of King Cotton* (1956), compares "cotton classing" or "cotton sampling" with tea sipping in that the tea sipper makes his living by virtue of the exquisite sensitivity of his palate while the cotton classer makes his through his clearness of eye and the sensitivity of his fingertips. Cotton is classed in order of sensitivity as Middling Fair, Strict Good Middling, Good Middling, Strict Middling, Middling, Strict Low Middling, Low Middling, Strict Good Ordinary and Ordinary.

This brings to mind a story told the writer by a University of Texas classmate about one Henry Ward, whose father was known all over North Texas for his expertise in judging cotton. This so-called Black boy, born outside the family pale, was the only child to inherit the father's unique gift. After his father died, it was Henry who had to be called if the quality of a bale were in dispute.

No Walnut Creek or Gilliland community would produce a cotton sampler with Henry Ward's expertise, but short of that, Brack Barr, his investment in his gin doubled, knew as much about cotton fibres and cotton commerce as any man in Texas.

16

Aunt Ann and the China Closet Theory

God made them, high and lowly
And ordered each estate:
The rich man in his castle,
The beggar at the gate.
—Old hymn no longer popular in Christian churches.

"Mr. *Jourdan's* little girl! Why, I didn't know they *had* a little girl!" The child heard it so often she felt herself to be a phenomenon, like the first bale of cotton in June that always stood on a sidewalk in front of an Austin bank tagged with its producer's name. Then, too, Aunt Julia said she was Eddie reincarnated, which added to her distinction, whether she knew what it meant or not. The day would come for her to realize that when anyone so small and insignificant was lifted to a table at a picnic or barbecue, it was to keep her from underfoot of the crowd; but at the time it made her feel embarrassingly important, what with passersby oohing! and ahing! over her to Mama and Papa.

From the time she could talk, which was early, she demanded and was rarely denied privileges belonging not to her age, but to her generation, which set her apart, in her mind, from age peers who were only second cousins, or, at best, firsts once-removed. Some of the firsts, like Frances Priscilla Fisher and the older Alfords, might actually, so far as age went, have been her grandparents. Nevertheless, she called them all without the prefix of "Cousin" if her brothers did. This privilege of title-dropping extended to young ladies and young gentlemen who were John's and Billie's friends, for John began taking her about with him so soon as she could sit up without being held.

Mama sent her along—in fluffy white for summer, with an eyelet embroidery sunbonnet; and in blue wool for winter, with a quilted silk hood. Hating anything on her head, she would rather have had any color than blue, especially pink, but Mama had a down on pink.

Mama's Cousin Augustina Costa (an elderly Sneed granddaughter who had lived abroad) once said to her, "Jody, pink makes your children look yellow." It may have been spoken carelessly in a bilious Texas summer when the boys were infants but it had sealed the child's fate so far as color was concerned.[1]

She might feast her eyes as much as she liked on damsels in delicately tinted hoop skirts dancing with gaily caparisoned gallants on the lace-edged fans young ladies passed back to her in church—it kept her still while sermons thundered over her head harmless as bumblebees blundering in and out through the open door—but she knew she'd as well ask for the moon painted on the scenes intriguing her as for a dress other than blue or white.

When John took her calling and the young ladies wanted to keep her overnight he let her stay. He could depend on her company manners—how could she behave less than properly when the Rowzee twins or the Misses Boyce or her cousins Alice and Laura took her on walks to look for baby rabbits and birds' nests, taught her the names of wild flowers, let her smell perfumes and Florida water in their bedrooms and watch in their kitchens as they stirred up teacakes and "kisses," then "lick the mixing bowl,"—that is, scrape the bowls clean of delicious dough and lick the spoons?[2]

In their dining rooms she could look at, though not touch, beautiful china arranged on shelves above everyday plates and cups and saucers, some of it at the very top so old and fine it was taken down but once a year just to be washed. These pieces had histories. It was the same in everyone's house including her own so that she came to ask over and over for the stories of butter dishes and soup tureens that had come to Texas in covered wagons while Indians were still around. There was one marvelous story: the story of how Mattie Austin's grandfather, coming alone with his family across the frontier, was suddenly surrounded by hostiles. Satisfied it was the end, at least for him, he stood aside with his wife and children, gun in hand, watching the war party plunder the wagon—he would shoot only at the

---

1. Cousin Tina (Augustina) was the maiden daughter of Jody's cousin Eliza Sneed and Judge J. B. Costa (born in Genoa, Italy) who was U. S. consul to Nice, Italy, 1857-1860. Reference: "Brown's Annals of Travis County and the City of Austin," mms. Austin Public Library.

2. Florida water: a precursor of bath salts; kisses: confections made of egg whites beaten stiff, sweetened and baked like drop cookies; nondescripts: a non-sweet pastry made to use the extra egg yellows—it was rolled thin, cut into long strips, then cut diagonally across into four or five inch lengths so that it curled into odd shapes when baked.

last moment, selling his life as dearly as possible—the others would be carried away captive. Then—a seeming miracle! From their possessions being piled on the prairie dropped a blue-eyed pink-cheeked doll! The pillage stopped midway—the doll was placed upright at a little distance and round it the painted warriors executed not a war dance but a devout ritual in a worship of their own, after which they departed, taking only the doll.[3]

In another and different lifetime—or so it seemed—on a sunny afternoon in Saltillo, Coahuila, Mexico, where her own young lady daughter was in summer school, it was fully borne in on her why she was so privileged and so popular in her babyhood. They, with several other *Unastadistas,* were staying in the home of a *señora* who had just acquired a polka-dotted cocker spaniel puppy.[4] The girls were taking the air on the front steps and playing with the puppy when a car full of strange young men came slowly back around the block, stopped, and asked to take the puppy for a spin. The *señora,* pleased, granted the request. Saltillo is a small town and it soon became known that the boys were driving all over, thrusting the puppy's head, its black-spotted ears flopping, over the car door each time they passed another group of well-chaperoned *señoritas.* Every exhibit drew a chorus of aahing! oohing! and *mira*-ing! and the slow-moving car moved slower for an eye exchange and a waved *Adios*!

When the little creature was at last returned to its delighted owner, the *Tejana* who had been "Mr. Jourdan's little girl" thought in surprised comprehension: Why, I was just like that puppy![5] Bait for John's girl friends, some of whom said he was the handsomest in Travis County!

Her birthday was the 3rd of April. She was almost three the first week in March, 1901, when he took her with him to the W____'s house. The W____'s were renting Uncle Zack's place on Walnut and John wanted to interest them in buying a piano. Twenty-one his past

---

3. Mattie was a cousin of the William Boyce girls whose mother, "Miss Kate," had been Mary Ophelia Stone. Known from infancy as "Kate" because of her babyhood demands for "cake" and more cake, she grew up on the Colorado where Samuel Stone operated the ferry and a nucleus of other businesses. She is buried in Austin's State Cemetery because her father was a veteran of the War of 1812 and she the capital city's last "true daughter." See Appendix J, "Stone's Ferry."

4. *Unastadiastas:* Citizens of the United States.

5. *Tejana:* Texan—woman. (*Tejano:* Texan—man).

birthday, he was trying his hand at salesmanship—anything to escape the tyranny of cotton.

While he talked pianos, she wandered from the room and down the hall toward the kitchen. She was not surprised to have a man gather her up, carry her and sit down in a rocking chair with her on his lap. She was used to that kind of attention. What surprised and puzzled her was that the man's head and arms were bandaged. Too well brought up to ask an adult a personal question, she sat perfectly still, not answering when he tried to get her to talk; teasing her with the age-old gimmick that the cat had got her tongue. It was no cat. Added to the mystery of the bandages, which meant the man must be hurt, was Aunt Ann's turning from something on the stove and saying disapprovingly, to two women drinking coffee at the kitchen table, "Miss L____, this chile ought not to be here!" The women, not answering, just stared at *her* sitting silent on the man's lap.

Why ought she not to be there, she wondered—had she done something wrong? If so, she couldn't think what. Aunt Ann, whom she looked upon as her very own mammy, hadn't even spoken to *her*. She knew, young as she was, that Black people weren't really mammies any more. It was just that Aunt Ann had stayed with her and Mama and taken care of them when she was born. Never before in her brief existence had she heard a Black person speak like that to a white person! It had somehow to do with her own behavior, but how? The longer she sat there with no one but the man taking notice of her, the more she was mystified.

John didn't come to the kitchen. He called her as he left the front room, giving her reason to slide from the man's lap and run back through the hall. Disappointed that the W____'s were not going to buy a piano, he was as silent as she on their way home. He sent her into the house with scarcely a word when he went to unhitch the horse. It didn't matter. Surely Mama could solve her puzzle.

Mama only half listened at first, then straightened to attention, asking questions. When she called John in to question him, he could tell her nothing. He hadn't seen the man—hadn't been in the kitchen—had known Aunt Ann was there and supposed the child was with her. Mama sent him to look for Papa. She and Papa had grown up on plantations. When she told Papa that Aunt Ann had looked round at the child in a bandaged man's lap and had spoken reprovingly to white folks about her being there, they both knew "something was going on" and that the "something" had to do with their little daughter. They must know at once what that something was.

Aunt Ann lived across The Field with her son, Commodore Jesse, in "your Aunt Mary's house." People died; but no matter how long ago or how many tenants moved in and out, houses that had been built for them continued to be theirs in name. Papa and Mama waited till Aunt Ann was sure to be home from making dewberry preserves or whatever she had been doing at the W____'s, then Papa hitched Queen, Mama's buggy mare, and brought the buggy round for them to drive over and find out what she had meant by what she said.

Aunt Ann's relief at their coming and her concern for her own position in the community were both evident in her greeting; nevertheless, there was no doubt as to where her loyalty lay. Mr. W____'s brother had come up the country on a visit and had broken out with smallpox, hence the bandages on his arms and head. The doctor had tacked a red card on the house but the W____'s tore it down as his buggy rolled out of sight.

The meaning of what was said on the way home was clear enough to one who would be three in less than a month, though she could not have put it into words of her own. It was to the effect that the W____'s had let John, in all innocence, walk her right into a deadly danger because they had no intention of bearing a stigma alone. If people, particularly those among the high and mighty landlords, wanted to come to their house and get smallpox, then no one could look down on them as the only carriers of the plague. It would be one way of leveling the social scale. So *that* was why she shouldn't even have entered their door, much less sat on the bandaged man's lap in their kitchen!

Papa dropped her and Mama at the yard gate and drove straight on to Manor. He returned with Dr. Richmond, who vaccinated the entire household from the scab on somebody else's scar, because that was the way they did it in those days. Later, he would ask for John's scab, it was such a good one, so that he could make more vaccine. She herself acquired a neat little rosette on her upper left arm, making her immune for years. She would be forty before another innoculation "took." But though time might dim her scar, nothing was to dim her memory of having been "exposed" and that just when Ethel Cato had come from Miles City to visit at Rose Cottage. Ethel, aged eighteen, eager for the parties being given in her honor, had no intention of flirting with the smallpox or putting up with the inconvenience of a vaccination.[6] Mere sight of Uncle John's buggy on the road sent her scurrying to isolation in the Chic Sales edifice at the back. The Gileses

6. John and Eugene had already given their parties for her.

were not alarmed, having all been vaccinated during a scare-up some years past.

Neighborhood fear of an epidemic blew over with but one other case reported, and it a month before, on the William Boyce place over toward Gilliland.[7]

In years to come, the drama of the occasion would serve to highlight the teaching of social theory in regard to a fallacy the cotton people seemed to accept as a life principle. This was a false application of Aristotle's classification system disguised under the maxim "a place for everything and everything in its place"; to wit, the notion that God had made the world exactly as it was in that point in time, and had assigned every person in it a specific, fixed niche like the dishes in everybody's china cabinets—the whole to remain forever so for the production and marketing of cotton.

7. *Texas State Democrat,* Sunday, February 2, 1901.

THE SCALPING—Ode to the Middle Buster or "Sod Buster."

The great rains roared—
They roared, they poured,
They ran the black soil
Down to the river—
Down those mile-long rows
The farmers plowed
That they were so proud
Of—right down
To the river;
And the rich black humus
Was stripped from the land—
Forever.

—M. Atkinson

TWENTY-FIVE year old Eugene "cowboying" in front of Rose Cottage.

17

The Aftermath

But pleasures are like poppies spread,
You seize the flower, its bloom is dead.
—Robert Burns, "Tam O'Shanter"

Eugene and John having attained their majorities, the goal of both was to escape the plow. John, lacking any luck in salesmanship, went to New Orleans to live with Baldy Alford and his wife, Katie. There would be a place for him in the railway mail service as there had been for Baldy and his desire to ride the trains would be satisfied. Katie Smith from Gravier Street would teach him Creole cooking—she and her small daughter and baby son would make him feel at home.

Eugene's store, a success during the preceding boom season, had turned out differently in 1901. The year got off to a bad start, weather-wise and cropwise, and, though excitement over a possible epidemic turned out to be a flash in the pan; the outlook in the fields did not improve as the months passed.[1]

Still, it being common practice for local merchants to "carry" their customers, borrowing from the banks till the next cotton could be made and sold, both Gilesburg and Sprinkle let duebills mount on the desk spindles.

To the Giles' dismay, and despite an increasingly poor aspect along the cotton rows, the Dessau store, inherently a center for beer-drinking conviviality, reopened—why look on the dark side? Every farmer must expect a bad season now and then! A second body blow was the federal government's establishment of a rural free mail delivery, the RFD. When the carrier's buggy replaced many small postoffices, Sprinkle's status was unchanged but the Gilesburg sign had to come down.[2]

---

1. Death of Miss Josie Kirkland noted in *The Texas Democrat,* Austin, 1901.

2. Ledger No. 6, 1930-1932.

When the fall crop came in, Austin bankers and wholesalers cabbaged onto what little cotton was baled. The owner of a small farm was lucky to so much as pay his debts with nothing left over for his children's school clothes. Even landlords were hard up. Seth Thomas took his blacksmith shop up the road to Dessau where there was a panacea for the mulligrubs (the blues) after the country merchants mortgaged their properties to carry over tenant farmers' indebtedness and furnish them still another year's necessities—hoping the 1902 crop would make up the deficit. People watching a clear sky for promise of a cloud said cotton was certain to come back—it *had* to!

It was a forlorn hope. The corn, food for teams that must make the crop, shriveled and died under a pitiless sun without putting on a single nubbin. The cotton did little better. Eugene's shoestring credit with the wholesalers, already frazzled, broke. The last set of entries for his store, Ledger No. 3, June 12, 1902, lists the following items:[3]

Sugar	50¢	Snuff	5¢
Candy	5¢	Oil (kerosene)	20¢
Tob (tobacco)	10¢	Tob	50¢
Tob	10¢	Pease	50¢
Candy	10¢	Rice	25¢
Salmon	40¢	Tob	10¢
Lard	30¢	Sugar	25¢
Tob	5¢	Black Draught	25¢
Cider (vinegar?)	35¢	Candy	5¢
Soda Water	10¢	Suspenders	25¢

The novice entrepreneur was up to his ears in debt—something like $1,500—for "merchandise and money borrowed to furnish credit to Tom, Dick and Harry." At that, he was more fortunate than his neighbor, Mr. Barr. Before Christmas, typhoid reappeared. The Sprinkle well, like shallow-dug wells all over the area, was a catch-all for drainage, as proved each summer when the store's molasses barrel sprang a leak and drinking water tasted of sorghum.[4]

Nobody gave a thought to location of a well other than that a vein of water must be struck. Medical science was not that far along. Water was simply water. Will Jackson and his family had typhoid and recovered. On New Year's Day, 1903, Brack Barr was dead. Jim Gooden and Cal Jackson carried the word through a stunned community.[5] It was the first death to be recorded in the memory of "Mr.

3. Ledger No. 3, 1901, p. 149. Black Draught: a patent medicine. Pease: note archaic spelling.

4. Reminiscence to writer: Mrs. Onie Jackson Kruger, Austin, 1977.

5. Interview with Mable Rowzee Penick (Mrs. E. Payson Penick, Jr.)

Jourdan's little girl,'' its emphasis due in part to her father's astonished reaction, in part to uniqueness of the moment and, finally, to changes in the community life style that followed.

It was a freezing cold morning. Hog-killing was the day before. She and Papa, all wrapped up, were standing at the back of the buggy house watching buzzards squabble with a rude flapping of wings over offal spread along a scaffold nailed to the building, when a Black man came striding out of nowhere, looming gargantuan to her inches. The scene would repeat itself in her imagination years later when, in a study of comparative religions, she came on Persian fire worshipers—the Gherbers—who held vultures sacred to disposal of the dead. The ungainly birds ranged themselves on the rim of a round stone tower, waiting, till a priest closed his ritual at noon with a clap of his hands, calling them in to their feast on a ledge just below; after which, the denuded bones would drop to a central pit.

The buzzards nested on Papa's piece of cedar land, across Walnut. She had seen the young stalking about and knew they were white where adults were black silhouettes in a sunlit sky circling the prairie, casting about for death among the grasses. Watching a loner afloat on a sea of air surrounded suddenly by a host of wings flapping madly, she could be sure the loner had found that for which it searched and had communicated with its aerial community. When she learned their Mexican name, *zopilotes,* she no longer thought of them as ''buzzards.'' The Indian syllables had no actual meaning for her, but poesy in the sound suited the serenity of their questing flight, then the sudden flutter of excitement and a practical drop to earth. Forever after that New Year morning these experiences would seem to encapsulate the natural sequence of things—the more so, learning of the Parsees' ''towers of silence.''

Tillie moved into Austin after collecting (according to the grapevine) ten thousand dollars in life insurance. She took with her, at Tom Rowzee's request, his eighteen-year-old son, Montella, so that Montella, too, might escape the cotton fields.[6] When word got about that, had Brack lived, he would have been forced into bankruptcy and that she had eschewed his debts, eyebrows went higher than ever, but she was within her legal rights and she had teenagers and a pre-teen of her own to look after—a daughter and two sons. Will Jackson and Robert Edward Jarmon, who had been Barr's right hand man, took over the store and the gristmill; Jarmon would also operate the gin. The firm became Jarmon and Jackson.[7] Jarmon, as senior partner,

---

6. Mrs. Kruger.

7. Ledger No. 9, 1937.

moved into the gingerbread mansion and Will was no longer on a stipend of ten dollars a month (of which Ola saved eight), with findings from the store for the kitchen thrown in, together with the area's second best dwelling rent free.[8] Erasmus Sprinkle would now take his meals, when he chose to do so, at Ola's table. Through the grim years, Sprinkle clung to a new hope—a railroad was actually coming!

Eugene, aged twenty-three, went back to the plow, leaving his family to dispose of his remaining staples as best they might while he sought to keep himself out of bankruptcy. He tells of his ordeal (paraphrased) in the 1930-1932 Ledger:

> My venture in the country store business left me with a pair of mules and a miscellaneous assortment of agricultural implements taken in on duebills. I was in debt $1200 to $1500 so I rented the Alford farm Thomas had had and the Cato place (his Aunt Julia's inherited acres) and took over management of our own Old Jourdan Place. I sub-rented part of my extensive holding to John Perry on the halves and worked the rest myself with Tom Dean and family (col.) as hired labor. We made good crops in 1903.
>
> One concern to which I owed money, tried to force me into giving them a preference lien on everything I possessed but I refused to allow it. I made a transfer of my crops to father conditioned that he settle my obligations to himself and other creditors as funds became available. With this for a club, I convinced the gentleman threatening me that the best thing for him to do would be to withdraw his suit and take his turn with the rest. Finally he got nice and in the end received all that was coming to him though it kept me two years behind a team of mules.

Texas, as shown by its map, tiptoes toward the tropics. Hence, up through Mexico come scourges of the hot lands mingled with the dust that neither patrol nor barbed wire can prevent from crossing a political boundary into an integral part of the same arid country as *Mexico del Norte*. Before Eugene's period of redemption ended, he and everyone else on the Black Land was aware that the Big Blow of 1900 must have borne on its wings more than the salvation of a boom cotton crop. Unnoticed through a following winter's hibernation, minute colonies of insect aliens naturalized themselves amid brush and dried cotton stalks edging the fields. It was in the spring of 1901 that Professor Fred T. Malley, state entomologist at the Agricultural and Mechanical College, Bryan, sounded an alarm: the boll weevil,

---

8. Mrs. Kruger.

Anthonomus grandis, prime enemy of King Cotton, had begun a march across Texas!

Within weeks, the professor so familiarized himself with the habits and disposition of the tiny creature, a quarter-inch in length, as almost to personify it in his rhetoric. Whether it actually had come on the Big Blow, as was surmised, or had stolen a wetback ride in the clothes of transient pickers, or had simply crawled over an international bridge on its own, he was certain it had arrived in small number and its seasonal progeny had not amounted, when discovered, to more than 200 individuals to an infested acre. The great danger lay in its ability to proliferate.[9]

His plan of combat, as set forth in the May 25th issue of *The Texas Democrat,* published in Austin, is fascinating as one views it with an imaginative eye. He dubs it "pan-shaking," estimates its cost at a few cents an acre and proposes that children equipped with pie tins and covered containers be sent along the cotton rows to scout for "flared" or yellowed squares, on the cotton plants or dropped to the ground below. They should gather all such squares (the insignia of an infestation) into their containers, then gently shake the stalks to dislodge any remaining insects and add them to the "catch." So great was the appeal of this idea that in some places merchants offered a bounty, so much per hundred, for weevils captured and turned in by the pan-shakers.

Next year, the professor thought, should the weevil survive a second Texas winter (he does not suggest praying for an ice storm or two or three hard freezes) cotton farmers must get ahead of a spring migration by taking advantage of the enemy's sweet tooth with a trick worthy of a G. Gordon Liddy, or the inventor of the Trojan horse. Let each farmer plow and plant a few pre-season rows, spray their preliminary "stand" with a weak solution of sorghum syrup and arsenate of lead, then sit back and wait for a hungry horde, roused

---

9. He said a lady weevil was very particular about incubation of her eggs, laying each to hatch in its own cradle—the "square" or embryo boll that forms before a cotton plant's creamy-papery blossom drops away. The gummy wet pap that should mature into the "vegetable wool" that is King Cotton's crown of glory must thus become, instead, a perfect baby food for Mrs. Weevil's nursling, not only as larve, but continuing to sustain it through its adolescent or pupa stage, until, bursting its chrysalis bonds in less than a month, it appears as a full-fledged adult, ready to reproduce its kind. This repetitive ritual would allow for increase, at geometric ratio, of at least nine generations in semi-tropic weather. A second spring would find the invasion out of control.

from its winter's sleep by the first warmth of spring, to rush into the poisoned fields and commit genocide, victim of its own ferocious appetite.

Neither of the professor's proposals worked. Moreover, lured by the spread of the cotton realm, other insect aggressors arrived, among them the army worm and the pink boll worm; but none except the weevil, the first to come, would gain a place in song and story alongside *la cucuracha* (the cockroach).[10]

The autochthonus song, which is to say the song that sprang from the soil and so became a part of the Public Domain,[11] croons:

Fus' ah see boll weevil
He settin' on de squar';
Nex' time dat ah see 'im,
He got the fam'ly dar!

then, over and over, with the quality of a spiritual:

Gotta have a home!
Gotta have a home!

Thus is the weevil memorialized though the boom crop of 1900 is forgotten; moreover, the creature is philosophically lauded by Eugene in the 1942-1943 Ledger: "Probably the boll weevil did more to put Texas on a cash basis than anything else, for due to it, diversity of crops was born out of necessity."

His Uncle Val expresses a somewhat different feeling in a yellowed newspaper clipping (lacking date and name of paper) which describes a reunion of Confederate veterans "on the south side of Gilliland's Creek in a grove of beautiful elms on Wednesday, August 24th—an ideal midsummer day":[12]

---

10. Army worm: larva of the *Leucania unipuncta;* pink boll worm—larva of the *Heliothus armigera.*

11. Public Domain: originally Texas' public land, of which the XIT holdings were a part before the Capitol Syndicate Contract. Example: when an individual sought to copyright a musical version of "The Eyes of Texas," the song was declared a part of the Public Domain and the copyright not allowed.

12. Veterans present were Towns Gregg, H. C. Norwood, L. B. Giles of Galveston, W. L. Giles, L. B. and Bart Giles, Val C. Giles, Pat C. Barnes; also present were seven descendants. (The writer cannot identify the Gileses other than the brothers W. L. or "Dock" and Val, therefore assumes all the others to have been unrelated to this family.)

According to the Perpetual Calendar the only years after 1900 and before Val C. Giles' death in 1915 in which August 24th fell on a Wednesday, were 1904 and 1910. Entry for Eugene's diary on August 24, 1904: " . . . annual Confederate barbecue at Mr. A. Norwood's pasture. Had an A 1 time."

> The grove where these pleasant meetings are held once a year is in the cotton belt of Travis County and as far as you can see in several directions there is nothing but cotton, cotton, except here and there a few acres of corn or sugar cane (sorghum). But how different the green fields look from what they did in former years. Thousands and thousands of white and pink blossoms ought to adorn these ___ fields of luxuriant growth but little cotton bloom can you see. Mr. Boll Weevil has settled that matter. I passed by and rode through ten thousand acres of cotton during the short time I was in the country, and I never saw fifty cotton blooms on my entire route (from Austin).

The time was not ripe for the devastation Val Giles felt was at hand. There would be some cotton in 1904—and when the end did come, "Mr. Boll Weevil" would not be the primary cause, or Mrs. Boll Weevil either.

FAMILY DINNER PARTY at Rose Cottage, March 20, 1904—(l to r) Val C. Giles, Virgil Edrington, Dock Giles, Julia Cato, Mrs. Lou Barnhart Giles, Ocie Cato, Laura Giles, Jane Maxwell. Seated (l to r) Maggie Giles, "Mr. Jourdan's little girl," Alice Giles, Miss Kate Barnhart.

18

What the Katy Did

It is the boast of the Katy management that it thrives with, not on, the communities it serves.

—M.K.T. *Employees Magazine,*
December, 1912

On the day in 1902 when it was determined that the M. K. and T. (Missouri, Kansas and Texas) Railroad would add an extension, Granger to Austin, that should run through the old Jourdan field, Sprinkle was elated. If five miles away, on Gilliland, the H. and T. C. (Houston and Texas Central), a spur of the Southern Pacific, together with cotton, had made Manor a county center of commercial prosperity and social gaiety, why should not the "Katy" do the same for Sprinkle? As the year came to its end no thought of death was in the mind of Brack Barr, postmaster, notary of the public, money lender, proprietor of a general merchandise store, a gin, a gristmill, a blacksmith shop—even an investor at ten cents a share in smelly black stuff from a so-called "gusher" at Beaumont.[1]

His sudden passing did not affect the Katy's plans. According to the Memoirs (paraphrased):

> Sprinkle was right alongside the tracks. A town site was laid out on the Barr property and the Railway signed up to maintain a regular station with a station agent in charge and a section gang. A substantial station house, a cotton loading wharf and cattle loading pens were built.

---

1. News item, *The Texas State Democrat,* Saturday, May 11, 1901: "W. B. Barr returned last week from Beaumont where he has been inspecting and investing in oil stock. He also speaks of the lumber mill as the finest he ever saw or ever expects to see." *The Texas State Gazette and Business Directory* gives oil stocks at 10¢ a share.

While the track was under construction Jarmon and Jackson did a thriving business. Starting of the trains, other than a work train, on July 14, 1904, was commemorated by a grand free ride to a celebration at Katy Lake on Berry Creek in Williamson County. Trains from Granger and Austin picked up folks all along the way. On that day almost everybody went to Katy Lake.

Sprinkle was envisioned as a large town in the near future. I made application for and was accepted as station agent on a salary of twenty-five dollars a month with ten per cent commission on ticket sales and express business.

Unfortunately, 1904 was a dry year with poor cotton. The water supply was so low Mr. Jarmon had to have the Katy haul water in tank cars to operate the gin. We made only forty-six bales of cotton on our home place and sold for eight cents a pound.[2]

His Memoirs do not say how many bales were made on the Alford and Cato places he had under lease but according to this report cotton was not so bad off as his Uncle Val's newspaper story makes out.

The Giles family never failed to take advantage of any invention or convenience that came their way and the Katy was like a new toy. On October 22nd Alice came in on the 6 a.m. train from a visit to the Land cousins (removed from Corn Hill to California) and at 11 a.m. she went out again with a party en route to the World's Fair in St. Louis. A photograph of the group shows Anthony Chriswell and Herbert Yager, Mr. and Mrs. Jarmon; their nieces, Misses Movie Jarmon and Mamie Earnest; the Giles sisters and Eugene, waiting for the 11 a.m. to come in—the five shirt-waisted ladies with skirts just clearing the ground and hats looking, today, like antic airplane models poised for flight but skewered fast to their great knots of hair by long steel hatpins equally servicable for weapons of defense should need arise. Posing of the group contrasts Alice's petiteness with her sister's height for Laura stands little short of Eugene's six-foot three.

The other travelers (Misses Myrtle, Fay and Etta Larue from upper Walnut, and Will Bowman) were on the train coming out from Austin. Round trip fare, Sprinkle to St. Louis, was $17.15 with the agent's percentage $1.71. Eugene, refused a pass by the Railroad and mindful of the scanty cotton crop, had done some tall figuring which he carefully recorded in his ledger. Round trip tickets, Austin to Sprinkle, were 65¢. If the four boarding the train there would let him pay that much of their fare, then buy their tickets to St. Louis from him, his percentage added to that from the other eight tickets, would

---

2. Ledger No. 9, 1937, p. 64.

pay his way and give him $1.37 to boot. He had their tickets in his hand when the train pulled in.

Sprinkle to St. Louis was less than a twenty-four hour journey. None of the twelve was interested in putting out the money for a Pullman; in consequence, they were a bit dazed when they arrived, but they had made reservation for rooms-with-bath at fifty cents a day, meals at twenty-five cents and someone from the rooming house to meet them and pilot them to a rest and a general freshening up before they set out on their ten days of sightseeing.[3]

Back in New Rochelle, Virginia, a boy was doing penance. Frederick Stewart Jackson, sixteen years his brother Will's junior, had come out to Texas before the Katy roadbed was completed and was dignified as a driver of the store's supply wagon, on its regular trips to Austin. Perhaps the western frontier had not lived up to his expectation or maybe being away from home for the first time with Christmas approaching sparked his youthful psyche to a show-off effort in compensation for nostalgia. At any rate, as night came on and the dark deepened, Will Jackson paced the floor, growing more and more concerned over the whereabouts of the holiday cargo, until a meteoric glory flashed down hill to Walnut, and up again into Sprinkle, accompanied by wild whoops and hal-loo-oo-oos from young Fred and still younger Ernie Barr. Aware of their lateness in leaving Austin, they were shooting the mules with Roman candles to make up for lost time!

Exhausted from a seven mile run in terror of their lives, the team, heaving and snorting as they trembled against each other in their harness, would be stove up for days to come. And seventeen-year-old Fred was shipped back to New Rochelle to quieten Tillie's threatening him with indictment for corruption of a minor. Ernie, barely edging the teens, had been, according to her charge, illegally plied with the season's cheer.

But Sprinkle was not to be too long denied the pleasure of the company of this last Virginia migrant. By the time Eugene was home from St. Louis and contemplating further travel, Will Jackson had taken his family, including a son, nine-year-old Elmo, back to his old home in the East where he listened with a softening ear to Fred's promises that never, never again would he induldge in high jinks if allowed to return to Texas—promises so solemnly and faithfully kept, even after the "Rat Hole" was established down by the creek bridge to give the Dessau beer dispensary competition, that one wonders if

---

3. Ledger No. 12, 1942-1943, p. 162.

THE CATO FAMILY, Miles City, Montana—standing (l to r) Leo, Myrtle, Ethel. Seated (l to r) Ocie Cato, John, Julia.

OFF FOR ST. LOUIS, 1904—(l to r) Anthony Criswell, Ed Jarmon, Mrs. Beulah Earnest Jarmon, Laura Giles, Eugene Giles, Novie Jarmon, and Mamie Earnest.

Tillie's tirade was entirely justified.[4] After all, celebrating the Nativity with fireworks is a perfectly good, Christian tradition, handed down to the South and Southwest by Spanish explorers who had it from those talented Infidels, the Moors, who knew just about everything there was to know about physics, chemistry and the creation of combustibles.[5]

By May of 1905, when the Katy decided to close its Sprinkle station for the summer, Eugene had a vacation plan figured out with as much attention to detail as he had given the St. Louis trip. He asked for, and this time received, passes for himself and also for Laura, hers as far as Nocona where he would leave her visiting with the Alford sisters, Jennie, Julie and Mollie.[6]

The Katy carried him from Nocona to Wichita Falls, whence the Ft. Worth and Denver took him to Dalhart, in the northwest corner of the Panhandle, where his objective was a ranch established by the Aaron Boyce family, recently of Gilliland Creek. The Boyces had filed on twenty sections (12,800 acres) of school land inside the XIT fences after the bad year of 1901—and after Imogene married the ex-ranger, Ira Aten, of Round Rock, who left the Ranger force to become high sheriff of Castro County with the prospect of succeeding her father's cousin, Albert G. Boyce, Sr., as general manager of the XIT. Aaron Boyce, a man sentimental about his children, was heard to declare that he could not have borne living so far away as Manor from Imogene.[7]

---

4. Reminiscence, Mrs. Onie Jackson Kruger, Austin, 1977.

5. Northerners seem always astounded by the celebration of Christmas with fireworks below the Mason-Dixon Line. Due to residential congestion and the loss of many homes due to flying sparks, cities now ban fireworks except for municipally controlled displays. Moors (Moslems) were "infidels" to Christians; Christians were infidels to Moslems. Infidel: one not of the true faith. Nevertheless, the Christian Southwest has adopted much of the Moorish culture—evidenced everywhere by intricately embroidered cowboy boots to be had in Nocona and Fort Worth.

6. Ledger No. 12, 1942-1943, p. 184.

7. In the hearing of the writer, as a child, John Jourdan, Sr. opined that a more cogent reason underlay the family exodus. He siad: "The Mexicans would have killed Ira Aten sooner or later, had he not left the Ranger force and Travis County." Aten's detachment had been more or less whitewashed at a Laredo hearing for cutting down on a Texas *hidalgo,* his son and servants, who were discovered crossing the Brush Country, far off but in range of high-power rifles. One member of the party was killed, another injured. The Rangers' defense was that they were after a party of horse thieves and if these *were* the wanted marauders they must have escaped into the brush had there been no shooting. Or so the story was told—a version that may not exactly coincide

John Boyce, the eldest son, met Eugene at Dalhart and they continued twenty miles via the Rock Island train to Middlewater, where they spent the night at the Middlewater division of the XIT, and next morning picked up horses for a twelve-mile ride to the Boyce ranch house. The family then consisted of parents and three sons: John, Dave and Ben, and two daughters: Belle (Isabelle) and Lucile. A third daughter, Lizzie (Elizabeth) was married to Harvey Avery. In a minor way it was a Gilliland Academy reunion.

There was, Eugene continues in the Memoirs, not a house between Dalhart and the Boyce ranch other than the railway station and the XIT headquarters at Middlewater—an isolation he was able to make further note of on a ride over into New Mexico. His visit was concluded when the Boyces arranged for him to go along as "cowboy helper" with twenty carloads of two-year-old heifers, part of a shipment of 10,000 head of cattle destined for his Uncle Ocie on the XIT's finishing ranch in Montana. His adventures as a "cowpoke" are here given as nearly as possible in his own words:

> I left Dalhart with a man named Crowell . . . Being a tenderfoot, never on a cattle ranch before, I was content to let Crowell be the boss, even to hogging all of the $25 paid for making the trip. So I bought myself some overalls and a prod pole and was ready to learn how. And I did!
>
> We were routed over Rock Island, Northwestern and Northern Pacific fast freight, meaning the local trains "went in the hole" while we whizzed through at top speed. Our first feeding stop was Harrington, Kansas, where we stayed overnight at a typical railroad eating and rooming house in the basement of which booze was openly sold—so much for the boasted prohibition of Kansas!
>
> I contented myself with drinking a bottle of Budweiser but "headman" Crowell stocked up on Old Crow Whiskey and stayed so drunk I had to take over checking the cattle and keeping him in the caboose till he sobered up. Our business was to patrol the car and keep the cattle standing. Occasionally one would get down and the prod pole, an oak staff ten or twelve feet long, properly manipulated from outside, would usually get them up.

---

with court findings in the subsequent inquiry.

An item in *The Texas State Democrat,* May 11, 1901, says: "Mr. (Aaron) Boyce has sold out his homestead of 330 acres for the handsome sum of forty dollars per acre. Mr. Boyce is one of the oldest and best citizens that Travis County ever had. He was here in the early days while the Indians were yet in possession of our country."

KATY STATION, Sprinkle.

"THE KATY"

M. K. & T. RY. OF TEXAS

Invites you to attend Free Excursion to

Katy Lake Park

Thursday, July 14th, 1904.

Special trains leave Austin 9:00 A. M. and 2:00 P. M.

PRESENT THIS CARD TO CONDUCTOR. FOR TIME CARD SEE OTHER SIDE.

SPECIAL TIME CARD. JULY 14th, 1904.

READ DOWN		STATIONS	READ UP		
A. M.	P. M.		A. M.	P. M.	P. M.
9:00	2:00	Leave..... .Austin.......Arrive	11:40	9:30	10:20
9:40	2:40	Leave ..Sprinkle......Arrive	11:02	8:50	9:40
10:00	3:00	Leave ...Pflugerville....Arrive	10:42	8:30	9:20
10:25	3:25	Leave......Burkland......Arrive	10:25	8:10	9:00
10:55	3:50	Leave ..Georgetown....Arrive	10:00	7:45	8:35
11:10	4:05	Arrive.. KATY LAKE ..Leave		7:30	
	6:30	Leave... KATY LAKE..Arrive	9:45	2:45	8:20
11:20	6:40	Leave .. Weir........Arrive	9:35	2:35	8:10
11:55	7:10	Arrive......Granger......Leave	9:00	2:00	7:40
A. M.	P. M.		A. M.	P. M.	P. M.

EXCURSION TICKETS—first train on Sprinkle track.

> Otherwise, it was an inside job for two men.
>
> There was no intimacy between Crowell and myself. At our next overnight stop we had separate rooms . . . I supposed he had given instructions to the hotel clerk to call us in time to make our train but when we reached the freight yard our train was gone. However, another XIT train was in with a couple of fine fellows . . . who readily took me on as part of their crew. We overtook my train at the next over-night stop, Jamestown, North Dakota. From there it continued to Glendive, Montana, where Crowell and I turned the cattle over to my uncle and his cowboys. A delegation of Indians was there to receive the dead ones and got two of ours that had died between Jamestown and Glendive.
>
> From Glendive I went to Miles City where I spent a month with the Catos. Miles City was then a wide open cattle town though I saw no evidence of drunkenness or disorderly conduct . . I took dinner one day at an XIT round up: chuck wagon and everything else for a cow camp setting. I also visited Camp Keough and other places of interest on my homeward way and spent a few days with the Boyces again; then home and back to my depot agent job at Sprinkle.
>
> My first observation of women and girls riding man-wise had been in Montana, but with so many of them constantly in the saddle, I had soon accepted this as the proper way to ride a horse. By contrast, when I was in Austin some months later and a local girl rode down Congress Avenue à *la masculine,* the spectacle attracted as much curious attention as a circus parade.[8]

Eugene was at home in time to stand best man with two couples on August 2nd: with his twin cousin, John Jourdan, and Dona Kimbro Boyce, early in the afternoon, and, later, with Gete (Eugene Edward) Jourdan and Alvah Mae Horton who was the daughter of Cary Quin Horton, an Austin civil engineer and bridge builder. John and Dona (fifth daughter of Aaron Boyce's brother William) were Gilliland Academy exes like her relatives he had left behind on the Panhandle ranch. Closely associated all their lives, these two had been more or less childhood sweethearts.

Faced suddenly with so much marital bliss, Eugene, after seeing the couple aboard the train for Beaumont where John was then stationed, began to wonder what, exactly, fate had in keeping for him. The first intimation came from the firm of Jarmon and Jackson. It appeared that he might acquire Jarmon's interest in the store and the gristmill, become junior partner and add the position of postmaster to

---

8. Ledger No. 12, 1942-1943, p. 165.

his position of station agent. Jarmon would retain management of the gin.[9]

Thus it was that on New Year's Eve, 1906, the sign across the Sprinkle storefront was repainted to read:

JACKSON AND GILES

and Ola became mistress of the gingerbread mansion where the junior partner would take his meals and sleep in the room with Fred when convenient. He could feel the more at home behind the counter because rather recently the store had bought and moved over his own old fixtures from Gilesburg.[10] Added to this, his Uncle Val's daughter, Annie B., who was his age, lacking a year, would teach the Sprinkle school (1905-1906) and also be Ola's roomer and boarder.

---

9. Ledger No. 9, 1937, p. 63.

10. Ledger No. 6, p. 53.

FRED AND ONIE Jackson in the Sprinkle store supply wagon.

PART FOUR

The New Grass

To every thing there is a season, and a time to every purpose under heaven.

—Ecclesiastes 9, 11.

19

Maid of Cotton

If cotton is King, then surely cottonseed is the Crown Prince . . .

—*The Story of Cotton* (brochure) Anderson, Clayton and Company, Houston, 1934.

It was late in time (compared with other states) when cottonseed, Cinderella of the cotton industry, came into its own along Walnut and Gilliland. Until an oil and seed cake mill was established alongside the Houston and Texas Central tracks in Brenham in 1891 ginners demanded that seed, except for what they could use to stoke their engines, be hauled off their premises. Simultaneously with the Brenham mill, seed houses appeared by the tracks at Manor and cotton wagons began pulling up to the depot and the scales for seed to be weighed and stored.

With gentlemen buyers bidding for the raw stuff at five to seven dollars a ton, instead of vast gray mounds smoking on the prairie where it was got rid of before its sticky, sweet decay became offensive, temper flare-ups enlivened Manor's streets. As when Mr. Allison whacked Mr. Cosey with his walking cane and stamped right on round to the sheriff's office to pay his fine for assaulting a competitor![1]

By 1906, ginning was concentrated at the Sprinkle gin; the Manor gin, owned and operated by Zackie's brothers-in-law, J. L. and Willis (Dub) J. Daugherty; the Nehring gin at Dessau that had replaced the Dederich gin and the Andrew Payton gin that had succeeded George Washington Cazine's gin at Fiskville. The Payton gin was on the road (now formally named for it) running through Amanda's edge-of-the-mountains field above Little Walnut.[2] All or most neighborhood gins

---

1. Interview, June, 1978, with Mattie Lou Allison Harris (Mrs. E. I. Harris), Austin, Texas.

2. On highway signs the name "Payton" is misspelled, using an "e" instead of an "a."

THE "Maid of Cotton"—insets Edith L., and the Waco Cotton Palace.

had burned: Aaron Boyce's, C. C. Baker's at Gregg's Corner; J. C. Maxwell's; and as so had an earlier seed mill at Brenham.[3]

Easy to ignite, difficult to extinguish, cotton seems to have an affinity with fire. A conflagration illumining the night of January 19, 1895, between Austin and Dallas, reduced the Texas Cotton Palace to ashes six weeks after coronation of Cotton as King had been celebrated by a jubilee lasting from November 8 to December 6—when Waco was declared the cotton capital of the South. The palace, imposing and wondrously hung with great plaques—shields of all the states done in bright-dyed cotton seeds—had been lifted from the prairie in less than six months. Given the haste of its construction, a fault of wiring may have been overlooked in the cupola where the fire started.[4]

Wacoans, thwarted in their expectation of continuing the jubilee as an annual event, were quick to substitute a May Day carnival with queen and floats.[5] Participating in this event was Edith L____, creator of an autochthonous Maid of Cotton—Ariadne, the classic spinner—an exhibit that escaped the fire.[6] "Mr. Dub," then in his thirtieth year and supposedly engaged to a hometown girl, was among the carnival visitors. He brought Edith back to Manor with him and they set her work of art on its low pedestal in their front hall. Sophisticated twenty-year-old Mrs. Daugherty was a welcome asset to Manor's young crowd. Her home was always open to them, her ideas fresh and generously shared.[7] They delighted in her, though they had not the worldly experience to gauge her thoughts as day after day she watched from her windows a monotony of cotton wagons rattling over rutted streets in dry weather, splashing mud and water when it rained.[8]

---

3. *Texas Almanac,* 1945-1946, "U. S. Prices of Cotton, 1790-1944." Interview, June, 1978 (prices of cottonseed) with Paul J. Lemm, Jr., Brenham, Texas, successor to his father, Paul J. Lemm, Sr., as manager of the Brenham Oil Mill till it closed in 1976. Brenham is the county seat of Washington County which had 40,000 acres in the "cash crop" in 1900 and had, in 1978, but 400 acres in cotton.

4. Barnes, *The Texas Cotton Palace,* (1964).

5. Writer's note: In this predominantly Baptist area, "Carnival," associated with Lent, was dropped the next year, though the annual May Fete has continued. Floats: wheeled vehicles, draped, commemorative of the marriage of Venice to the Adriatic—early water carnival.

6. Autochthonous: sprung from the soil.

7. Reminiscence, Mrs. Harris, 1977. Pictures of Edith and her Ariadne are among the Giles Papers.

8. Imogene Kuempel Acklin (Mrs. J. E. Acklin), Round Rock, Texas, who is compiling a scrapbook record of the Tuesday Reading Club, has a picture of

In December of '96 a Daugherty boy was born. The baby's advent put no damper on Edith's gaiety nor lessened her popularity with the young folk. In September, when her son was nine months old, she announced her intention of going back to Waco for a visit. She would take the midnight train on the 16th which would be *Diez y Seis*—Mexico's Independence Day.

Manor and the surrounding country had a considerable Spanish-speaking population. The *Diez y Seis* celebration was in full swing with Anglos looking on at the dancing when Edith's friends left the Fiesta to see her and her child off on their trip. The train whistling away into the midnight was the last they were ever to see of her, her baby or the Maid of Cotton. Eventually they would hear, through the grapevine, of Mr. Dub's soft drawl admonishing her brother in a Waco courtroom to "Put that gun away—it may go off and hurt somebody!" and that she was bringing the boy up under her own name instead of as a Daugherty. A long while later they would learn from a news item that she was the wife of a titled Englishman in California. Mr. Dub would then be the father of two daughters—the younger's mother, sister to the elder's mother (killed in a train wreck)—their family well known on Gilliland.

Less than a month after Edith left Manor, two of the town's young ladies organized, with eighteen others, the Tuesday Reading Club which would meet in rotation at their homes.[9] At the second meeting one, then another, having dropped out, the Misses Giles were invited to replace them. Travel from over on Walnut may have been a bit hazardous but was hardly considered fraught with mortal danger. Nevertheless, then as now, a good story was a good story. A year, lacking a month, from the day of Alice and Laura's membership, as they neared Dessau and Harris Branch (a Gilliland tributary) on their return from a meeting in the William Gregg home, their buggy mare, Maud, spooked by only she knew what, tipped the buggy off the bridge. They hung suspended, while Maud, her fright momentary, stood sedately above till two Manor gentlemen appeared on horseback. Helped, uninjured, to regain their dignity, and their buggy put back on the road, they would read in the next issue of *The Texas State Democrat* of their miraculous escape from "almost certain death."[10]

Some years, some movements, moments, incidents, are pivotal in

---

Manor at this sloppy stage.

9. Mrs. Acklin's scrapbook, "Minutes of Meetings."

10. *Texas State Democrat,* Austin, Texas, October 10.

an area's cultural development. Influence of the Tuesday Book Club, by no means the first social reach for something beyond an accustomed pioneer hospitality on Gilliland, would be lasting, as witness its 80th anniversary in 1977.

On Walnut, Eugene, before his store venture, sponsored by his sisters, had thought to organize a Pickwick Club to function in members' homes. The membership list included, among others, the William Boyce girls: Beatrice, Claudia, Dona Kimbro and their brother, "Kie," who was known formally as Albert G. Boyce (a favorite name in the Boyce clan); John Jourdan, Jr.; The Misses Almquest, Mary and Joanna; Aaron Boyce's daughter, Belle, and sons Ben and Dave; Eula and Dora Payton; the older Yager girls, May, Kitty (Katherine), Annie and their brother, Herbert; and May and Claud's friends, Robert and Owen Smith. It was a mixture of ages, with half a generation between eldest and youngest, and also a mixture of scruples that failed to blend. For instance, Methodists danced—at least some Methodists, and Baptists did not dance, at least most Baptists; hence, there being no agreement on a program of activities, the Pickwick Club died a-borning.[11]

Although that idea fell through, impetus of the boom crop in 1900 and the Katy Railroad in 1904 had given Sprinkle a considerable status among the county's smaller cotton towns in 1906 when Eugene became a store partner. The year began in sadness; Erasmus Sprinkle, aged ninety-four, was dying. He had been for some time bedridden at his grandson Hayward's house where his daughter Cremona came and stayed to nurse him. A month after his passing on the 25th of January, Cremona Barr died, exhausted by her labor of love.[12] He was buried in the Gregg Cemetery on Gregg Lane with his grandson Braxton, Braxton's wife, Tillie and their baby boy, Aubry.

Father and daughter had seen what they started as a roadside commissary rise through several major steps to an extent of twenty-three houses centered round a merchandise store, a gristmill, a blacksmith shop, and a gin. The first step, in 1884, had been designation as a federal post office; the second, in 1901, relocation of the Burditt school; the third, in 1904, arrival of the railroad. Monies raised by neighborhood subscription in 1900 built the new schoolhouse which had the overall aspect of a church though not tacked to heaven by a spire like the Baptist Church at Manor and the Methodist Chapel on Little Walnut, the latter an outgrowth of Fiskville's Methodist

11. Ledger No. 9, 1937.

12. Interview, June, 1978, with Onie Jackson Kruger (Mrs. Herman Kruger), Austin, Texas.

"camp meetings."[13] Its modest bell tower lacked even a bell and rope. Dependence was on a hand bell rung from the doorstep to call into its one room pupils of ages seven to eighteen in grades first to ninth—those among the forty enrollees who stayed out of the cotton to come to class.[14]

Church services were held on occasion—as in the Old Burditt Schoolhouse—when someone brought a preacher for a day or a theological student was invited from Austin for a practice sermon. Sprinkle people, barring the Black folk who were Baptists or Methodists, were mostly of the Progressive Christian Church persuasion, and, like some Gilliland Creek Methodists, inclined to take religion lightly so long as no one was actually sacrilegious; that is, outside of a little blasphemy when aggravation was sufficient and ladies absent. Passengers looking from train windows, hearing the conductor announcing, *"Sss-prr-ĭn-kulle!!!"* as sonorously as if it were a grand metropolis, asked sometimes, "Do they call it that because of that little church over there?"[15]

They couldn't have been more wrong. Had it been a church, baptism would have been by immersion.[16] It was "Sprinkle" because it was the mark a tall old Virginian with a white beard and intensely blue eyes had left on the map of Travis County. Yet none of the young ladies and gentlemen or the younger boys and girls who frolicked on the lawn of the gingerbread mansion in 1906 would have supposed for an instant that before the century was out that mark would have been erased by smoke, the smoke of progress, as surely as the Texas Cotton Palace at Waco had gone up in actual smoke and flame.

---

13. Ledger No. 9, 1937; essays, mms. by Eugene Victor Giles (in old tablets) "Jourdan Park and Its Pioneer Background, February 24, 1956" and "Some Rambling Thoughts on Now and Then, January 26, 1957."

14. Each spring the state requires a head count of all children of school age in each school district, regardless of what school, public or private, a child attends. By this count the head tax is distributed for the coming year among public schools.

15. Writer's reminiscence, from playing around the Katy station in childhood.

16. Alexander Campbell's followers immerse, as do Baptists; Methodists (the writer has been told) immerse only on personal request. See fn. 4, Chapter Nine, "Miss Sallie's Boy." Congregations in the so-called Christian Church, like Baptist congregations, have complete autonomy; hence, the tendency to spawn deviate offshoots. Alexander Campbell's disciples lost no time in dividing on principles due to interpretations of Bible readings. The

distinction of Progressives and Non-Progressives (Church of Christ) was and still is based largely on the use or non-use of instrumental music in worship. The latter groups, also called Firm Foundationists and Fundamentalists, came to be known in the upper portion of the Black Land cotton realm as "Square Bale" Christians, and "Round Bale" Christians after Anderson and Clayton developed the more compact round bale for replacement of the bulky square bale in shipping. A few years back when the sects held state-wide conventions in Houston simultaneously—but not conjointly—a reporter on one of the major papers, meaning to be folksy and writing "by ear" distinguished the groups as "Round *Bell* Christians" and "Square *Bell* Christians."

ALICE AND OLD MAUD.

20

Uncle Jim's Horseless Carriage

Do you remember that first automobile,
Cranked on the side,
How merrily you went and with what pride—
Till the dern thing sneezed and coughed and died?

—Eugene Victor Giles (n.d.)

Ola Jackson had long run her smaller home as a sort of hostelry. Under her management, the gingerbread mansion with the lovely Marechal Niel rose climbing the spindled lace fans of the veranda, built to be a social center, became more so than ever with two eligible young men—Eugene, her husband's business partner, and Fred, his brother and chief clerk—in residence.[1] Grading of the Cameron Road (though that counted for little in wet weather) swelled the number of friends driving over from Gilliland or out from Austin on pleasant Sunday afternoons. Where better for young ladies of the community and their house-partying guests to gather and spend the lazy summer hours?

At the same time, a wave of movement sweeping the county's landed gentry into Austin was beginning to be felt. Some said it was due to the public school system Seaborn Sneed, Jody's cousin from over at St. Elmo on Onion Creek, had helped to effect.[2] Concentration in Austin schools allowed for many advantages one-room, one-teacher rural schools could not provide. To avail one's children of these advantages required residence within city limits. The result, at first unnoticed, would be absentee landlordism with the county population, both colored and white, reduced to tenantry, most of it a great influx

---

1. This was the cream, not the old-fashioned yellow Marechal Niel rose.

2. Seaborn G. Sneed was Travis County school superintendent, 1893-1894.

THE GINGERBREAD MANSION.

THE SPRINKLE SCHOOLHOUSE.

of share croppers from areas east of the Black Land.[3]

In the summer of 1905, John and Jody bought, for $1500 cash, a two-story house on a block-long graveled driveway islanded with elm trees and directly across from Zackie's barn and horse lot which took up the south side of a place at 809 West 22nd Street which he had acquired after a year or two of commuting between Walnut Creek and the courthouse as county treasurer.[4]

Jody took possession of the house at 805 West 21½ Street in September when the school term started, although her daughter had been enrolled at Sprinkle in the April head count for that school district.[5] Since she could already read and write and knew the multiplication tables, at least the easy ones, having been taught those things at home, "Mr. Jourdan's little girl" was taken by the first grade teacher at Wooldridge Elementary, Miss Eula Hill (who was Uncle Jim Maxwell's cousin Lou Maxwell Hill's daughter) to the second grade room and there entered on a new life-style which was to be a dichotomy, much after the fashion of *hacienda* folk in South American countries today.

Papa, in his fifties, who had no use for "living out of paper sacks," as he described the patronage of grocery stores, would "keep batch" with Billie (whom he had never in his son's life addressed or spoken of as other than "William") during the week. Friday morning he would drive into town, returning to the country with his wife and daughter in the afternoon. Driving them back Sunday or early Monday morning always found him eager to be "out home" again. He would maintain Sprinkle as his address and voting precinct.

He had some time since retired from active farm management, turning it over to William, then twenty-three, and to Henry Neans, a neighbor's son about the same age, who moved, a loner, into the head tenant's house. Which did not mean he retired into idleness. To avoid the "paper sack routine," he supervised production of foodstuff in his mother's plantation way, seeing to it that the place provided everything for the two kitchens other than sugar, coffee and white flour. With Billie an expert cook, neither household suffered at meals

---

3. Ledger No. 6, 1930-1932. Up to this time, most local cotton growers owned their own acreage.

4. *Austin City Directory,* 1895-1899.

5. When the north side of the block was built up with residences and the address was corrected to 809 West 21st Street, the trees were removed and the street was graded during or soon after World War I. Pupils are enrolled where ever they are in April for the coming school year if they will be of school age in September.

for there was never any dearth of fruits and vegetables, either fresh or preserved, fresh or smoked meats, milk, butter, eggs and breads baked from choice corn or wheat grown specially for the purpose and ground at the Dane's gristmill on Little Walnut after the Sprinkle mill went out of business. Even lye soap for cleaning purposes was made in a washpot over an open fire, cut in chunks and spread to dry on boards in the smokehouse.

He served as consultant, sometimes as banker, for the younger men. The three were a satisfactory human combine.

The entire family summered in the country and in one of those summers the blue dress edict came to an end, for the vacationer marched up to the counter in the Sprinkle store to buy with her own nickels and dimes a length of pink seersucker. Quite horrid in retrospect, it was the best the store offered. Made into the simplest possible frock, it was worn till it fell in shreds. The subject was never again mentioned—nor was there ever another blue dress.

The wearer of the drab though beloved pink, always lonely in the country, found just one drawback to the situation in town. Uncle Zack and Aunt Belle's colored cook's room was between stable and kitchen; her name was not Mary, but their first cook had been a Mary. No memory lapse, only sheer self-indulgence made all cooks thereafter Marys so that when Uncle Zack wanted this one, it was both an inconvenience and an embarrassment to have a stentorian shout for "Mary!" resound from corner to corner of the block—how was one to tell if it were his cook or his schoolgirl niece he wanted?

Wooldridge was a no-nonsense school. Miss Mary Johnson, the prinicpal, and Frank Strand, the janitor, were euqally conscious of their relative positions in authority.[6] Frank, a tall, past-middle-age, brown man, not only kept buildings and grounds clean (one brought him a nickel next day after spilling ink and the spot disappeared), but also kept guide bells ringing on time and indirectly supervised conduct just by his presence. One knew Frank would report a flagrant indiscretion.

There were three hundred pupils and a playground divided between boys and girls, with a division line neither overstepped. Frank rang the five-minute-bell to cease games, get drinks and be ready for class to take up, both at arrival time and at big and little recess. The bell allowed for queues to form and all to quiet down before teachers supervising the grounds gave the sign to march into the building

---

6. Paragraphs in this chapter relating to Wooldridge School and to the family are from the writer's recall.

where room teachers waited at their doors to see that seating was accomplished with no undue talking. When the last bell rang, all talking ceased.

Since there were seven grades, there were seven rooms. Each room was divided into "low" and "high" sections. One passed from low to high at the middle of the year, but in the second term low sections were mostly filled with those whose birthdays had made them too late to enter in September. If they "caught up," their report cards would allow them to skip a half term when the next year began. Lows recited one thirty minutes of the hour, while the highs studied, and vice versa for the other half—a routine under the home room teacher except for arrival once a week of the art teacher, who taught water coloring, or, alternately, the music teacher, who taught reading and writing of musical notation, and singing. Each taught the class all together for a whole period.

There were no prayers—everyone was well aware of the separation of church and state amendment in the Constitution as it applied to schools, although no one doubted the teachers' being God-fearing, or their ability to instill that fear without mention of the Deity. No assemblies were held; sometimes an after-school play or pageant was put on, to which other classes were invited. The schedule was easy to follow because one knew, every minute, what one was expected to do. The subjects stressed were grammar, geography and arithmetic, with special emphasis on reading and deportment.

A hush-hush excitement pervaded the house in town in the late spring of 1906. Dona, young John Jourdan's wife, had to be smuggled out of Beaumont in a mail car—a matter to be kept forever secret, first because it was against the law for an unofficial person to ride in a mail car, and, second, because Beaumont was closed up tighter than a drum to contain a yellow fever epidemic and breaking quarantine was also against the law. It was carefully explained to the child that both laws had to be kept for necessary reasons, except in the most extraordinary circumstances, one of which was that Dona could not be left alone, a stranger in a strange town, with an epidemic on, when she had not as yet been exposed to danger. Modesty had kept her indoors before the alarm was sounded and now, when her husband was not allowed to go across town to her, she had to be got back home by hook or by crook.

It was all very wonderful and mysterious, especially to come home from spending the day with a friend and find a brand new baby in the house. The baby's name was Nell, which seemed inadequate for such a marvelous morsel of humanity, but the baby's mother refused to add

SUNDAY GUESTS AT Sprinkle—"Gete" and Alvah Mae Horton, Alvah's sister, Marie, Fred Jackson, and Billie Jourdan.

SUNDAY AFTERNOON at the Gingerbread Mansion—Onie Jackson with dog.

a further distinction. Some years later the little one and her young aunt would seek to remedy the lack by adding "Eugenia" in honor of their favorite cousin, Eugene, but after a brief interval the idea was abandoned.

Across the graveled driveway, the house beyond the horses' stable and the cook's house had also been blessed with a baby—Z. P. Junior, a trifle older than Nell. The first of Zackie's six grandchildren (three boys, then three girls), young Z. P. would be known to his family his life long as "Brother" just as the first girl (young Alvah) would be known as "Sister." The other boys would be Cary Quin and Eugene Edward, Jr. (Demp); the younger girls would be Jannice and Hazel Jane.

Out at Sprinkle, Eugene's Uncle Val's daughter, Annie B. ("B" for Barnhart, her mother's maiden name), at her father's and her Uncle Dock's solicitation of the school board (Thomas Jefferson Rowzee, Robert Edward Jarmon and William Maxwell Jackson), was elected teacher for the 1906-1907 term. That year she, too, lived in the gingerbread mansion, paying out of her salary of fifteen dollars (or was it twenty-five? She could not remember which in 1975) a month six dollars for meals and lodging with one dollar recovered by giving Onie piano lessons.[7] Several years older than the Rowzee twins, then twenty, who were always in any group decorating the veranda along with the roses, playing croquet or being "kodaked" on the lawn, "Miss Annie B.," a town girl and a home girl, rarely stayed for weekend fun. Although catching the Katy every Friday evening to be met by her father in his buggy and returned to the station on Sunday, to be in time for school Monday, diminished her income by sixty-five cents each trip, she felt the break in routine to be worth it, having soon learned that gentle manners and a knowledge of subject matter, though good, were not sufficient of themselves to hold interest and keep order in a classroom.

At home where her mother, her Barnhart grandmother and her spinster aunts, Miss Annie and Miss Kate, owned almost the entire 700 block between West 22nd and West 23rd Streets, a scant three blocks from The University of Texas, preparations were under way for a new development. Her brother, Val C., Jr., married at Christmas of 1904 to Minnie Greenhill, was back at home. With "Miss Min" was her sister, "Miss Ed" (Eddie Mae Greenhill). Given Miss Ed to preside over the dining room, Mrs. Lou Giles was about to open *The Subur-*

---

7. Miss Annie B. Giles' experience with the Sprinkle school is from interviews with her, and with Mrs. Onie Jackson Kruger, in 1975 and 1976.

ban, offering home facilities to university girls and to her daughter release from an irksome treadmill. It would be at *The Suburban* that Miss Annie B. would celebrate the end of academic responsibility by inviting Onie and the small girl from around the corner, who had entered Wooldridge the previous September, to a formal dinner followed by a performance at the new institution downtown—a moving picture show. One remembers one's first movie. This one, not particularly edifying, was called "The Mother-in-Law."[8] Though silent, it was full length, unlike skits the child had seen at the amusement park on lake Pontchartrain two years before when she and Mama spent a month with John and the Baldy Alfords in New Orleans. Onie, approaching sixteen, was due to enter Whitis, Austin's boarding school that catered to daughters of well-to-do farmers, ranchers and out-of-town business men in conjunction with a day school for Austin girls who had not made it, or didn't care to, in the public schools.

Spring was over on Walnut and Gilliland and in Austin by the time winter ice broke up in the Northwest to float down the Yellowstone River past Miles City where Ocie Cato, sheriff of Custer County, was about to be elected state senator. A cattleman become a living legend from Alaska to Mexico (they said he was so knowledgeable on the trail he could tell a cow's color by looking at her tracks), he nevertheless had lost that which he treasured above money and above glory: the darling fourteen-year-old, Leo, was gone with a heart attack.

This was the year (1907) when Uncle Jim and Aunt Jane were back in the *Austin Directory:* this time at 809 West 23rd Street, and Amanda was renting a house at the corner of West 21½ and San Gabriel in order that Dora be treated for incipient tuberculosis. Eugene records nine uncles and aunts "in shouting distance of each other," not to mention his cousins, the Charles Alfords on 20th Street directly across from the university campus. Charles, on the city council, would be elected and reelected until he was dubbed "the dean of councilmen."

Only Dock Giles and his children and the two Payton families, John's and Ephraim's, were left in The Field where Katy trains whistled and tooted and slid noiselessly down or stalled amid the cotton on their way up what was said to be the steepest grade between Austin and Kansas City. The kinnery talked among themselves and doubted the Gileses would ever move to town on account of the shadow cast by Flora and Maggie.

---

8. This was an early silent movie.

In 1908, the Katy made Sprinkle a telegraph station, requiring a professional telegrapher, for which position Eugene did not care to qualify. As a matter of fact, he was beginning to question his entire Sprinkle relationship, officially speaking, for progress had brought its inevitable curse: competition. Quoting from the Memoirs:

> Another blacksmith shop was opened by a man named Reeves and J. A. McWilliams built a house on one of the town lots, a small box business house on another and opened a meat market, later converting it into a general merchandise store.[9]

Before too long, Jim and Jennie reversed the family trend into Austin by moving back to their two-story house on the edge of The Field next to Brack Barr's embryonic township. Childless, they had no need to bother about schools, nor, having acquired five farms and a hundred thousand dollars in cash, had they any need to concern themselves further about business, whatever cotton might do, and cotton was not going well. Boll weevils were ravenous, the weather continually uncertain. It was as if the terrible Galveston storm had unsettled the balance of nature.

Living nearest the Maxwells in The Field, John Payton and Lura had a son, Billie, and four daughters. Last of the four was the baby, Pauline. This small child, sitting in her Uncle Jim's lap, parting his white beard, attempting to find a place on his face to kiss him, brought back long-buried memories to James Corry Maxwell. He called the child "Polly" and told her, "I loved a little girl once, back in Virginia, named Polly," and when the child, thinking of Aunt Jane, asked "What became of her?" he said, "She died." When the small one sighed in sympathy and leaned back against him, he finished the story: "I thought then I would never marry, but after awhile I decided that I would, because then I might have a little girl of my own, like my Polly."

Jane, coming to the door, listening, unable to give her Jim that which he most wanted, said, "Well, now you've got your little Polly!" and went back into the kitchen. Pauline would be thirteen when he died. Her aunt would call her from the Sprinkle school and he would go, holding her hand, murmuring "Polly."[10]

Pauline was five or thereabout when her Uncle Jim, reaching three score and ten, bought his first "horseless carriage." Neighbors might have called it "Maxwell's folly" had it not been that he was

---

9. Ledger No. 9, 1937.

10. Reminiscence, Mrs. Pauline Payton Johnson.

perfectly able to indulge his fancies even though cotton, up in price, was declining in output. Cotton men in general were content to argue philosophically, as they blued the air with pipe and cigar smoke or chewed and spat over a veranda rail, that fool machines like that thing in Jim Maxwell's yard would come to nothing, their arguments generally ending with the clincher that even if such things could be made to work, *women* would never be able to handle an automobile! In an affluent area where ladies had their own rigs and themselves harnessed and unharnessed favorites like the Misses Giles' old Kate or Jody's Queen (only Jody never did) such a thought was not only absurd, it was unseemly—something like the idea of the side saddle's giving way to riding astride. Still, though mentioned only with covert smiles, the ideas were in the air.

But James Maxwell, despite his age, was a forward looker, long ahead of most of his venerable peers. He moved, in 1909, to dispense with the hunting hospitality that permitted friends and friends' friends to drive out from Austin for a day's shoot on Walnut and adjoining prairie pastures. He hoped to save what remained of the quail and dove population. Posters were made and signed by twenty-three land owners—Negro, Anglo, Swedish, absentee and resident—banning gunning on their properties.[11] The move split the community in twain. Reactionaries resounded with fury; chief among them Walter Sanderson, who had Zackie's Walnut Creek place under lease. Owning no acres, this sportsman had recently bought a hundred dollar rifle with gold mountings. Any birds coming under the fence onto the land he controlled, he said, would then belong to him and he soon announced with gusto that he had "got every one of Old Man Maxwell's quail!"

This public boast roused Jane to verbal fireworks unbecoming to a lady, which in turn brought sister-in-law Belle (Mrs. Zackie) out to dine and do a little soft-soaping in the interest of family peace and harmony. Although Zackie hadn't signed the proclamation, she said, he certainly felt that "something should be done to preserve the game birds." And so the movement fell by the wayside. Zackie knew his politics.

Those who represented the opposing stance, unable to attack on any other grounds, said that the Sandersons, two unmarried sisters (Sarah Bertie almost a generation the younger) and a bachelor brother bound to one another by a deathbed promise to their departed parents (who had come late to Texas from Alabama), were going to wake up

11. One of the posters is among the Giles Papers.

one day and find themselves in a hard row of stumps. This because, having no heirs, they had consistently rejected advice to become landowners. Instead of a homestead, they were putting their investment in fine teams and fine tools with the cultivation of cotton on rented acres their mainstay. Traditional Deep South Southerners, their feeling was that the right of a gentleman to hunt on the acres he controlled should never be abridged.

The horseless carriage brought as little satisfaction as the effort at bird conservation. It sat in the front yard, an object of curiosity to passersby on the Sprinkle Cut-Off Road linking the Cameron-Dessau spur and the Cameron Road proper. It looked more or less like a buggy. It had to be cranked up by one handle to start it to sputtering, and it had to be guided in the way it should go by another handle. Its owner made no attempt to drive it. This nephew or that with a mechanical bent would drop in to take Uncle Jim for a ride into The Field or around the corner to the post office. If he or Aunt Jane had business in Austin or need for a few days' inspection of one of their farms, the one at Menard, for instance, they rolled along as usual behind a horse they could trust to take them where they were going and bring them home again without breaking down and having to be tinkered with. When Pauline's grown-up brother Billie was the driver, she sat beside him in her Uncle Jim's lap, her brief legs sticking straight out in delicious horror lest they touch some of the unfamiliar gear and cause a wreck, ending up with her entangled in the chain that ran along the side of the machine.[12]

It was on November 11, 1909, that the Maxwells astonished the world with a joint birthday celebration (he sixty-nine on the 11th, she sixty-two on the 10th) in the form of a barbecue on Walnut to honor five remaining Maxwell ex-slaves and all descendants of "Maxwell people." Some two hundred attended the day-long occasion. The special honorees were Fannie Allen and her brother, Moses Fraction; Tom Madison; Jack Black and Amy Black Shanks; all of whom were presented cabinet-size pictures of several different groups, and everyone else a souvenir mug. Newspaper accounts caught the attention of relatives and friends over the United States.[13]

No one thought at the time that a rather forward little eight-year-old girl, Juanita Shanks, who was William and Amy's granddaughter and the daughter of Dave Shanks, a Prairie View graduate, would one day be a political power and repeatedly a presidential guest at the

12. Reminiscence, Mrs. Pauline Payton Johnson.

13. See Appendix K.

White House; or that she would end up in her seventies on the Dallas City Council with a park dedicated to her.[14] Even less, perhaps, was any cotton man thinking of an era to be called Black Gold. Their only uses in the early 1900s for the stuff gushing from a salt dome near Beaumont was to spread it round their houses when there was sickness, or on stagnant water to limit mosquito breeding. The idea of a mosquito as a bearer of malaria or the dread yellow fever was but months older than the news of Spindletop and no more accepted by everybody than the farfetched notion that transportation would one day be accomplished without recourse to animal traction so that barnyards, except for cow lots, must disappear.

Still one had to admit that times were changing, what with talk of the main roads being macadamized so even rubber-tired vehicles could roll over them with ease.[15] Already creeks were bridged, except the shallow ford where Cameron Road crossed Little Walnut just below the baptismal hole which was used mostly by Black Baptists. This ritual facility was directly under windows of the White folk's Methodist Chapel with its white steeple pointing heavenward, on the west side, and next, on the east side, to the Dane's wind-powered gristmill.[16] The juxtaposition seemed odd since Methodists immersed but rarely—and only by special request.

A kingfisher flashed up blue against the green woods lining the creek banks each time a vehicle rolled down to the water, and, unfailingly, splash of the wheels sent Jody into a thin, fearful wailing.[17] Perhaps it brought back the terror of a five-year-old's long, frightening coach ride among strangers, down from Comanche Town on the Indian frontier, where her mother was dead, to an unknown home in the judge's house on Onion.

What was not admitted by the general public was how uneasily King Cotton's crown was now resting on the hoary locks. The righteous, Black and White alike, felt that outcome of their great gamble was in the hands of the Lord; and in the Lord's own good time

14. Juanita Shanks Craft (Mrs. J. E. Craft), Dallas, Texas. The Juanita Craft Recreational Center, 4500 Spring, Dallas, was dedicated in 1977.

15. Macadamized: topped with crushed stone; system named for its originator, John Loudon McAdam (1756-1836).

16. Jens Jensen, according to his daughter, Mrs. Mary Jensen Pils, Austin, brought his family from near Copenhagen to Travis County in 1905. He owned land on both sides of the creek and it was in his pasture that convicts were encamped to work the Cameron Road. The writer recalls passing a chain gang in zebra-striped vestments, the leg-irons clanking as they labored.

17. Writer's recall.

''Cotton would come back''—it must, or His people would be ruined, and surely, He would not let His people down. Meanwhile, country merchants restocked their shelves by going ever deeper in debt to the wholesalers in town, making it possible for the ''havvies'' duebills to mount higher and higher on their own desk spindles.

At Rose Cottage, New Years Day found Alice sickening of a malady diagnosed as ''female complaint.'' Some days she felt better, some worse. Her illness increased through the srping. So did the burden of a helpless household that was coming to rest entirely on Laura as their father faded into a day-dream of the past, a condition doctors ascribed to ''softening of the brain'' (now known as ''paresis'').

On Valentine's Day, 1910, Eugene went into Austin and handed in his resignation as Sprinkle postmaster, recommending that William M. Jackson replace him. Due, he recorded in his diary, ''to underhanded machinations on the part of local Republicans'' the appointment went, instead, to the merchant who had zeroed in when Sprinkle's progress, thanks to what the Katy did, was at its height, thus dividing a mercantile business just when King Cotton's throne had begun to shake. The notation further reads:

> Will be very glad when I am relieved of all public responsibility—the simple life—the good ole farm for me from now on.

Later he added:

> My thirtieth birthday—can hardly realize it. Once made resolution I must be worth $30,000 and 30 years old before I married. Well, I'm 30, all right.

On Easter Sunday, March 17, Alice died in the Austin City Hospital. The diagnosis remained uncertain. On the 27th they would have celebrated her forty-third birthday. Instead, they took her body to their Aunt Lou's Suburban to lie in state through Monday evening and on Tuesday afternoon they buried her in the family cemetery on Walnut. They left the gold band that betrothed her to Lee Dunn where it was and they put on her feet the dainty pumps she had been too sick to wear after she bought them.[18] It was Jane who said to Laura, ''You didn't want her married and gone—now she's gone anyway.''

On October 13th, Eugene's wish for complete release from public

---

18. Writer's recall. Alice's death described in Eugene's Brown Diary No. 1, p. 228.

responsibility was effected through resolution of the Jackson and Giles partnership in the Sprinkle store and the senior partner was left to carry on alone. Thus the era of Black Gold dawned with Uncle Jim's horseless carriage sitting idle in his yard and King Cotton's subjects struggling to bolster a tottering empire, unaware that as masters of the plow they had been chief among the agents of its destruction.

The Indians could have told them, but the Indians were gone.

THE SPIRIT BRIDE

> In the name of our Lord, I . . . in the form and manner wherein thou hast promised thyself unto me, do declare and affirm that I will one day bind and oblige myself unto thee, as my husband. And all that thou hast pledged unto me I promise to do and keep unto thee by the faith that is in me.
>
> —Christian Rite of Betrothal

THE ELDER siblings spend a day at Rose Cottage with Dock: Mrs. Land, Mrs. Henry, Val C. Sr.

21

The Search for New Horizons

Cotton is a frontier crop restlessly seeking new horizons.
—David L. Cohn, *The Life and Times of King Cotton.*

Eugene's Memoirs ascribe the beginning of the great cultural change he witnessed on Walnut and Gilliland to 1901, saying that no one recognized what was happening at the time. Actually, it had been happening all along, for, to coin a phrase in Biblical fashion: Out of the past cometh the future, the eternal evolution set in motion by the Creator.

By 1910, many enemies of the imperial reign of Cotton, beginning with the kingpin, Boll Weevil, had made themselves manifest and not every enemy was an all-devouring insect. Fritz Kruger, always ready to experiment, hearing of a wonderful, self-sustaining grass, sought to revivify a dried-out pasture by an imported planting.[1] As advertised, the stuff called "Johnson grass" was impervious to weather, hot or cold; to all creatures, great or small; even to human attack. It soon outran desired boundaries. Dropped sprigs rooted themselves where they fell; roots missed by the plow crept into turnrows and covered a field within a season, running underground; seeds, undigested by work animals, sprouted to be fertilized by their droppings. The Indians would have laughed to see Mother Earth, scalped of her native grasses by the sodbuster, growing an alien, unwanted crop of new hair.

Those of Frederic and Harriet's grandsons who had stayed on the land, not satisfied with the returns they were getting for their endeavors, were casting about for new horizons. Early in December of 1910, Eugene, with John Jourdan's son, Billie; Amanda's sons, Andrew and Bradford; and Steve Jarmon from Sprinkle, entrained for Mexico to look at land Andrew and his Aunt Jane and Uncle Jim had

---

1. Giles Essay in small tablet.

bought under the Diaz regime. It was near the line between the states of San Luis Potosi and Tamaulipas, not too far from the railroad that joined Tampico, port city of the latter, with the capital of the former. The exact point is difficult to determine, using Eugene's notes: after a train ride leaving Tampico at 3:30 in the afternoon of December 6th, they arrived in time for supper at a stop called Coco, a spot too small to be found on any available map. Thence they were driven by buckboard to the Penetis Ranch, which was within fifteen miles of the Maxwell and Payton holdings "on the little river . . . As fine land as I ever saw," Eugene eulogizes it, "deep grayish black, rich with the accumulation of rotting vegetation (and an) abundance of water in the river."[2]

From there they rode over to the "Atascador town site in the jungle which was just being cleared." They passed a field cleared in May that had been "planted with sticks and gone through with *machetes* (may-shāy-tays) to yield twenty to twenty-five bushels of corn the acre."

They pitched camp on Andrew's place where every day they caught sight of wild game, peeping from the jungle growth, giving promise of fresh meat to supplement the forty dollars' worth of supplies they had bought in Tampico. William was delegated to operate the improvised kitchen, but, try as they would, they were unable to augment the menu through their skill as huntsmen.

On the 12th, Eugene, William and Brad took a twelve mile morning climb to one of the few landmarks, cited in the Memoirs as Little Mountain, "which we scaled—the view from the summit is superb." In the afternoon, Eugene and Andrew went for a six-mile hike into the jungle growth. On the 16th, the party packed up, walked back to the Penetis Ranch and were driven by car to Coco to take a train that was bringing the Maxwells to visit their own stretch of bamboo, palms and cypress. Jane and Jim would spend the Christmas holiday there, Jane putting in most of her time fishing. There was little time for talk other than that the campers had missed all shots at deer and been frequently so covered with wood ticks they "had to scrape them off with a knife."

Despite the wondrous fertility of the soil, the trip led to no more "buys." Williams' opinion was: "There's going to be a war down there; if I bought land in Mexico I'd have to fight for it and I want no part of a war." Ragged, sandaled troops getting on and off trains in

---

2. Ledger No. 4, 1904-1905.

CONFEDERATE REUNION at Rose Cottage, ca. 1910—(l to r) as named by Eugene: Bill House, Jim Maxwell, Camp Giles (not related to Eugene), Val C. Giles, Lee Giles (not related to Eugene), Bill Dunson, Dock Giles, Virgil Edrington.

Monterrey and Ciudad Victoria had impressed him far more than the richness of the jungle floor.

The rightness of his view soon would be attested by postcard photographs in cut-rate drug store windows showing fly-blown corpses—a kind of news largely omitted from papers above the Rio Grande, inasmuch as tabloids had not yet invaded the Southwest.[3] For his aunt and uncle and his Cousin Andrew, upshot of the unpleasantness he foresaw would be reduction of their Diaz land grants to crumbling yellowed deeds doomed to the ash can.

Close of the year found the house at 805 West 21½ vacant, as it would stand for a decade. Underlying its abandonment was a malady not yet named in American dictionaries or discussed by educators studying the problems of school drop-outs: *dyslexia,* or dysfunctions in reading. The question of the moment was why should a child in the top echelon of her class at Wooldridge, The University of Texas teacher training school, find long division incomprehensible in the fourth grade? Why should anyone (including herself) suppose a child with superior distance vision, and an expert reader, could not clearly identify arithmetic symbols from 1 to 9 when the configurations were the least bit complicated and every digit must be clearly identified? There had never been a need to identify every letter in a word; one read by the shapes of words and sums had hitherto been simple enough to do in one's head.

Spin-off from this undiagnosed difficulty was another, rapidly growing into a horror: the realization that, having been "conditioned" into the fifth grade because she could not do long division in the fourth, if she did not work the condition off there, she would be given a failing report at the beginning of the summer. That would hold her back while others went on into the sixth in September. There was a single recourse—if one were too ill to be in school to finish the term, there could be no damning report card when school closed.

Perhaps her father sympathized with her plight whether or not he knew a cause for it. He had himself refused to attend school in a day when only the elderly were supposed to need glasses for reading and no one suspected his extreme nearsightedness. At any rate, when September came and the notion was presented that he and Jody put their problem daughter into the boarding school where Onie was, he acceded—which was to cost him a year's tuition for a few weeks of instruction not listed in the curriculum provided by the two maidenly ladies owning and operating the institution.

The new pupil was one of three "little girls" put off together in a

---

3. These exhibits, as the rest of this chapter, are from the writer's recall.

part of the house away from other boarders old enough to date. In this corner allotted to careful conservation of her innocence, she heard her first smutty story. Out on the grounds, if she happened to stand with one or both arms akimbo (a posture she was never thereafter able to assume unconsciously), she was chastened with the cognomen "teapot" or "sugar bowl" by one or another of the staff. In the dining room she learned by experience the truth of what she read later in a woman's magazine—that it was boresome to the taste buds to be served the identical menu on the same day of each week, day in and day out. In All Saints Chapel on Sunday mornings, she learned that one knelt for prayer, sat for instruction, stood for worship—a bit of ritual it was well enough for someone brought up in the austerity of Methodists, Baptists, Presbyterians and two kinds of Christian churches, to know.

In the living room, she found that being compelled to sit through Sunday afternoons surrounded by young ladies studying or writing letters or reading the Bible, but not talking, was worse than when, as a child, she had been taken visiting with Mama and had run into the house away from other children who teased her, and Mama said, "If they don't like you, why don't you stay in here with us. Maybe Mrs. Whoever-Was-Hostess will let you look through her stereopticon." The views of Niagara Falls and Washington's Monument and other like wonders of the world, distilled in sepia, had been better than being imprisoned with a staff member as jailer to see that nothing broke the monotony before brown paper-bag suppers were passed around.

One lesson to be most valuable throughout her life was presented, but not by a staff member. Maidie Dealey, from the Dallas publishing family, passing through a classroom empty except for Jody's daughter, paused to read what was being written on the younger girl's desk top with chalk. Surprised to find it was verse that rhymed, but limped, Maidie stayed to teach scansion, something that fell in line with the writing of music—all the youngster, who had never been able to carry a tune, had gained from the musical instruction at Wooldridge.

Jody could carry a tune. She could pick out tunes on the piano's black keys "by ear." It had been a long-standing feud between her and her daughter that the latter must learn to play the piano, whereas each time the child, to whom color was vital as food and drink, begged for art lessons (or a pink dress), the answer was always either that artists were disreputable people who starved in attics or: "Your Papa would never pay for them." Why Papa would pay for music lessons when he wouldn't for painting lessons, and even drive her into town to Miss Dot Thornton's house to be taught (until requested not to bring someone again who didn't care to learn) was a mystery equal to why long division remained incomprehensible. Somehow, the question seemed to go back to the forbidden pink dress. Nobody thought to test for a slight tone deafness that went with the dyslexia.

Whatever art and music might have to do with pink dresses, a young ladies' finishing school was plainly no place for a public school learner accustomed to competition with the academically ambitious. Again, the only out would be to sicken, as she had before. When she could no longer attend classes satisfactorily, though the sickness appeared to be in her mind (as Jody's sick spells often were), Papa came and took her away.

Jody made a last effort to determine her daughter's destiny. Amanda's daughter, Eula, was teaching the Sprinkle school that year. Well known, both as a sound academician, thanks to her university degree, and as a strict disciplinarian, she had been given a teaching assistant and a more munificent salary than Miss Annie B., Eula would live with them and drive Queen back and forth to Sprinkle.

With someone to coach long division at home so that the process was at last understood, that part of the problem quickly melted away, but there were other angles to the situation, Papa and Billie recognized, no matter if Jody had shut her eyes to them. Papa, coming on his daughter alone in an isolated corner of the yard, told her gravely what the future might hold for her through her making intimates of those with lesser opportunities than herself. He said nothing of boy friends, but she was older than she had been—on the edge of her teens. She caught his underlying meaning and agreed. She condemned herself to a loneliness that would be long, but not everlasting.

So, the penalty of humiliation Jody had imposed by sending her to finish the school year in a country school was lifted without necessity for rebellion.

22

After the Decline, the Fall

And he changeth the time and the seasons; he removeth kings and setteth up kings.

—Daniel 2:21

Looked at through hindsight, an outline of the future for the prairies between Walnut and Gilliland becomes clear by 1911. On Wednesday, February 13th, Eugene and Fred Jackson entered into a partnership to operate a dairy at the spring branch below Frederic and Harriet's old house. They would call it "The Live Oak Spring Dairy" for the giant live oak overshadowing their venture. Eugene's Uncle Jim Maxwell was financing the purchase of thirty-nine cows and a bull, three horses, some hogs, a cream separator and other equipment from Fred's uncle, Will Jackson.[1] That night, the two bachelors celebrated the deal and their joint birthdays—Eugene was thirty-one, Fred twenty-six—by going in town to the Hancock Opera House for a performance of "The Girl from Rectors."

On August 3rd, Eugene, together with Herman Kruger, who was batching and operating the place that once had been Ocie Cato's, across the road from the Gileses, attended a dance Onie gave at the gingerbread mansion before entering The University of Texas for another term.[2] Gaiety that evening was without thought of the occasion as a farewell party, but in December the "Barr Store" failed

1. Eugene's diary for 1911 is in twelve little black booklets, one for each month. It has this notation; but according to an instrument in the dairy files, the purchase involved a second lien, since Eugene had secured a first lien at 8% interest with his personal property: a horse, farm machinery and equipment, six head of mules and a lien on 1911 crops plus all future crops until paid, with interest to rise to 10% if suit were brought for collection. However, leeway was left for the debt to be paid ahead of time, if desired.

2. From the 1911 diary series.

under weight of the duebills accumulated on Will Jackson's desk, attesting to how often and how long he had grubstaked tenantry on the failing cotton farms. He and Ola, after a brief interval, moved into the Maxwell's two-story house where that aging couple was suffering isolation of childlessness and the handicap of Jim's ill health. They, Jacksons and Maxwells, would operate a small dairy on their own account. Elmo, in his first year at A & M, came home to aid his parents.

This change-over in one's family living arrangement was indicative of the total change coming in the social life of the countryside—a change to which was basic the triumph of Johnson grass over the soil, for where the abused earth no longer could provide a cotton yield sufficient for profit after the costs of production were totted up, cows could graze on the alien herbiage at no cost to the investor. Besides, Austin, advancing toward a population of twenty thousand, was ready for a creamery industry. With land prices dropping faster than they had risen in cotton's heyday, the scramble, for those who could read the signs of the times and had children coming on, was to get into town, take whatever work offered that would support an austere respectability, and strive to fit the next generation for a professional rather than an agrarian livelihood.

The greater tragedies were not to be for those—Blacks and White alike—who were "selling out" for what they could get and taking their chances in town, but for those clinging to their faith in cotton—determined that it would—it must—"come back!" Some would mortgage the land to keep the plows going; others, despite successive losses, would gamble their very lives on "a change in luck and the weather."

Three friends of Walter Sanderson—the landless entrepreneur with the finest gun among them, and, like the gun, the finest car—would go on his note for a few thousand to save him from bankruptcy. "Nobody," they would tell each other (Eugene was one of the three) "knows cotton like Walter; he has kept a record of every bale he ever made!" They were not really fooling themselves. It was a gesture of *noblesse oblige.* In the end they would have the note to pay and the Sandersons would disappear downriver with Walter, a security guard for another man's possessions, dying broken-hearted in 1942. Miss Ada preceded him by five years. Sarah Bertie who, about the time she moved into town, spent several days with the writer, faced oncoming years with $300 from the sale of household effects. She would manage as companion to one and another of Austin's aged and dying.

Shock must come to many with foreclosures on the land. Worse shock with penalties for liens given on non-existent teams, tools, cords of wood. When banks brought indictments, a gentleman's son, found guilty, would die after three months of penal servitude. William Shanks' tall son, Walter, looking down on the deputy sheriff he thought of as a Red Neck, when the latter approached him with handcuffs, pulling a gun, would jerk the gun away, turn the officer of the law over his knee, administer a drubbing, send the gun back by an aged friend and disappear from Travis County.

The future for Cotton's devotees on Walnut and Gilliland—those whose eyes stayed closed to signs and portents—was to be an agonal epoch, long drawn-out. By the 1930s cotton would be down to four cents a pound—twenty dollars a bale.

On John and Jody's place, William, age twenty-eight, to keep abreast of the weather and beat a dropping cotton market at its own game, was building a sixty-foot by thirty-foot galvanized iron storage shed, pillaring it with cedar posts, one set every ten feet, and shoring up the northwest corner for a ten-foot square granary. He would, he said, hold products, including cotton, "till the price was right." And he aimed, when this shed was full, to build another just like it. Brad Payton, two years younger than William, watching the sheathing go on, remarked, "He'll be a rich man—if he lives long enough!"[3]

The next year (1912), an automobile garaged in the southwest corner of the shed bore no resemblance to the contraption in Jim Maxwell's front yard, so fast had been the evolution of "those fool machines" that ran without animal traction. It had been demonstrated by a Mr. Barber from Austin one day when several branches of the family were assembled at Amanda's house. The question, "Would it pull?" brought a suggestion that it be tried out at the steep Bull Creek Crossing. "If it pulls that, John," Jim Maxwell advised his brother-in-law, "you better buy it."[4] It pulled the grade and was bought, but it would not pull Black Land mud. When rains set in, William had it towed to the Giles' gate onto the macadamized road and left it there till dry weather, leaving his horse waiting till he came back from town.

The car, a right-hand drive K-R-I-T, had nothing in common with the Maxwell's predecessor other than four wheels and a need to be cranked (being without a battery, it ran on a magneto). It required a powerful man to do the cranking, and he had to be careful lest it break his arm, it came on with such a kick. Otherwise, it was easy to

3. The writer also watched the shed building and heard this observation.

4. The writer rode with the try-out group and was later given driver's training by Mr. Barber.

handle, even by the fourteen-year-old daughter of the house though she never could get the hang of the "tin-lizzies"—primitive Fords—Eugene and Brad added to their vehicular equipment. Eventually, Jim and Jane would replace their exhibit with a handsome Buick any young relative could drive for them. Of the older generation, only Zackie would drive for himself.

When Texas woke to a new source of revenue and issued numbered license plates, the K-R-I-T was 772 in the state—a number the family felt had historical significance and supposed would be theirs forever, like a cattle brand!

On January 8, 1913, Herman, work dropped on a cottage he was building in anticipation of his marriage to Onie, wired Eugene (attending a short course in soil conservation at Bryan) that Will Jackson was dead of an apoplectic stroke; he had driven into town the night before to bring Onie to her father's bedside.[5] She did not return to her classes; instead they honeymooned in the new cottage as soon as it was ready.[6] Her mother and Elmo moved in with them before the year was out and the Jackson dairy herd was merged with the Kruger herd, leaving the Maxwells to begin a search for a satisfactory live-in nephew—an impossible task given Jane's increasingly irascible disposition.

In June of the same year, John Jourdan, Sr. (a portly and dignified five-foot-two beside his over-six-foot son, William), at his daughters' insistence, doffed his wide-brimmed black planter's hat for a grey Homburg, donned a new grey suit (but kept his black string tie as appropriate to his shortness of neck) and boarded the Katy Flyer with his nieces, Eula Payton and Laura Giles, for a visit to his sister Julie in Miles City.[7] "Cato," he was to say of the erstwhile cowboy, "has at least one child named for him in every family in Montana; and to every one of his namesakes, he has given a cow and a calf."[8] He

---

5. The writer walked with her father to the Maxwell home and was present through the night while it was hoped by the doctor that Mr. Jackson could be pulled through till morning when it would be possible to take him to a hospital and operate. He was buried in Gregg Cemetery.

6. The writer was present at this wedding.

7. Until John made this trip, both he and Jim Maxwell wore planter's hats in contrast with Zackie and Ocie, who had "been up the trail" and always wore rolled Stetsons. Dock Giles' distinctive headgear was always referred to in his family as "Pa's Darby"—not the Appalachian "Paw" of the comic books, but pronounced as the first syllable of "Papa."

8. To present a newborn with a cow-and-calf "to start a herd" was Texas custom. When the increase of this gift to the writer was sold, the money was

would be speaking then of O. C. Cato, state senator, a man of wealth and renown throughout the Pacific side of the continent—a lifelong Democrat whose personal popularity was such that, in a Republican state, he was being urged to put his name on the ballot for U. S. senator.[9] The six-week excursion ended with a tour of Yellowstone National Park, where the avuncular visitor from Texas became "Uncle John" to everyone in the touring crowd.[10]

A week before putting the group on the train, Eugene, William, and Eugene's father, again visiting Amanda, drove again to the mouth of Bull Creek, where, according to Eugene's diary, they "located a camp site which, as a company, we expect to purchase."[11] On the 11th of August, Eugene, William and Brad (the latter three months married to Eva Pearl Thorpe) met the Katy Flyer at Sprinkle to bring the travelers home. Two days later, while Eugene was at home sewing cotton sacks, a dark blue cloud came up from the west in mid-afternoon. From the cloud a white lightning bolt, striking straight down, felled William to the ground from atop a wagon as he lifted a pitchfork of hay in an effort to get the crop in ahead of the storm.[12]

The elderly mortician said, as he prepared the burial, "This is the most beautiful man's body I have ever seen." William had been thirty in May. Years later, Eugene, reading back over his diary, penciled in the margin by the Bull Creek campsite notation, "Pipe Dream."

John, Jr., with Dona and the small Nell, came from San Antonio and stayed until Christmas. Not long after they left, would-be buyers from Hutto offered the elder John $60,000 for his 300 acres, saying it was the finest farm in the Black Land—"Not a sprig of Johnson grass

---

always hers to spend (in consultation) with her parents, as she chose.

9. From family conversation, obituaries, and a political announcement in *The Democrat,* Miles City, Montana, October 5, 1912.

10. A clipping from the *Farm and Ranch* (Dallas) pasted without date in one of Laura Giles' scrapbooks (she was a frequent contributor to this paper under the pen name "Olga Lewis") begins:

> I'm seeing America first, for I've been from ocean to ocean, from the Gulf of Mexico to the Canadian border. I have crossed all our great mountain ranges—have been on summit after summit of those same snow-capped mountains that used to inspire me as a child; and now that I've been through Yellowstone National Park, I think I'm about ready to see Europe.

11. Eugene's diary—large brown book No. 1, 1913-1920.

12. Same citation as above. The writer did not witness the catastrophe but did see the lightning descent and still can call to mind the vivid stroke of white fire and the terrific thunder crash.

on it!'' When his daughter asked why the offer was refused (for he was alone then; even Henry Neans, married at last, had gone in search of a new horizon) he said he would not know where to invest the money.

On April 13th, 1915, Senator O. C. Cato arrived in Austin with his daughters, John and Myrtle, and Myrtle's small daughter, Mary Cato Williamson.[13] He had come south in the grip of heart seizures that began, spiritually if not physically, with Leo's death in 1907. Very soon his entire family gathered in an apartment at 816 West 12th Street, just above Pease Park, which ran along Shoal Creek. It was a wet spring after a prolonged drouth. Austin was built on a rise between Shoal and Waller creeks. Through a three-year drouth, no one had given any mind to accumulations of brush and other flotage, partly windblown, under the bridges. A torrent rushing down from the hills on the night of the 24th, dammed at these bridges, rose into adjacent houses. The Catos waded out, Ocie determinedly carrying Mary Cato, despite protests. In the morning of the twelfth day after, May 6th, he died at 1600 Nueces, and there, two days later, a memorial service was conducted for him with Masonic rites.[14]

Someone went out to find a few late bluebonnets to put under his hands. As the casket was carried down the walk, starting the journey to Miles City for a state burial, five-year-old Mary Cato and her two-year-old cousin, Charles Claiborn (Ethel's son) walking just behind, reached up, touching it. A grownup in the procession, noting their tiny tribute, said softly (was it Laura?), ''Uncle Ocie would have liked that.''

On February 2nd, 1917, the United States served diplomatic relations with Germany, and, on April 5th, declared war. Here or there on Walnut and Gilliland a youth was called up, uniformed and sent overseas to countries as alien as a valley of the moon today. There had been no radio or TV to alleviate the isolation of the Texas they grew up in. Elmo Jackson went, and Herman's brother, Will; but along those creeks, young men of the right age were few and far between—rather, the area abounded with genteel maiden ladies who sometimes had olden bachelor brothers. Such groups were held together by loyalty to family and pride of land—and an intolerance of outsiders: possible in-laws who might upset a delicate balance of affluence, influence and social status, and, more especially, force a division of the acreage. (A cultural anthopoligist's mind harks back to Egyptian Pharonic and Peruvian Incan systems of brother-sister

13. Same citation as fn. 11.

14. The writer was present at this service.

marriage to keep imperial powers intact; those, too, were cotton countries.)

On the morning of April 4th, 1918, James Corry Maxwell died; and just before dawn of July 22nd, Jane's brother, John, having started to rise, fell dead in his bed of a heart attack. Doubly bereft, she was frantic with anger and despair. Life had even denied her and Jim the joy of a golden wedding they had looked forward to for the preceding October as a fitting culmination to their years of companionship. Her painful bout with erysipelas at that time had left her perfectly bald. She sent the fallen hair to a cosmetoligist, who returned it in a wig appropriately styled for an as-yet-unknown but greying Shirley Temple. Post-Victorian Jane, plain of feature, but with all the dignity of a slight lordosis stance, took one look at the monstrosity and placed it on its little form in her curio cabinet, there to remain so long as she lived—after which it would be dropped, with the yellowed Mexican deeds, in an ash can.[15]

At the end of a year, loneliness and despair drove her back into town. Characteristic of her Hetty Green tendencies, Eugene's diary has the mundane statement November 5th, 1919, "I went to get Aunt Jane's milk cow—either to keep her for her feed or to buy her for $80."[16] More importantly, he sold a single bale of cotton for forty cents per pound, such was the temporary effect of the war on the imperial commodity which was about to give over, so far as Walnut and Gillialnd were concerned, to the plebiance of dairy products.

Jody and her daughter reopened the house on 21st Street immediately after John's death. They would not enter the country house again. Austin's growth was such that residences had replaced Zackie's horse lot and he had removed the three generations of his family to an imposing structure where he and O. C. Cato were developing a suburban area they called Plainview Heights. This development was a part of the Rutherford Plantation where, more than four decades agone, the milk wagon had spilled their Atkinson cousin and Mrs. Rutherford into Buttermilk Branch.[17] He provided Gete with a dairy herd and oc-

---

15. The writer well recalls the wig; also that, among themselves, it was not beyond the imaginations of adolescent peers to refer to austerely wound hairdos like Jane's as "cow-pile coiffures."

16. Same citation as fn. 11. There were those who said of people like the multimillionaire, Hetty Green, they "would skin a flea for its hide and tallow!"

17. Plainview Heights dealings are recorded, like all property changes of hands, in the county's Deed Books.

18. *Austin City Directory.*

cupied himself as treasurer-secretary of King's Sanitary Creamery in which he had an interest.[18] (When Austin became over-citified for his taste, calling him to task when he pulled into his accustomed parking place downtown, he turned his car and drove home, fuming, to call a taxi and return to his office where, in an hour or so, he recovered his usual good temper.

Suburban living had its advantages, but also its disadvantages. In the family's absences, their house twice burned to ashes, with every Austin fire truck standing helplessly by because there was no water pressure beyond city limits. The third time, the house went up on a less elaborate scale.

It was at the door of the original dwelling that Eugene and William knocked one afternoon, to have it opened by a small boy with two smaller brothers ranged supportively behind him. "Where," asked one of the visitors, "is your daddy?" "Daddy," said the child, not sure which of many relatives he was addressing and slightly awed by the towering height of whoever-it-was, "is still at the Creamery." "Oh," pursued the questioner (perfectly aware that "daddy" in that house was Zackie), "Well, where is your papa?" This was more difficult. Carey Quinn Horton did not live there. How was a grandchild to know where a civil engineer might be building a bridge? Came, after a moment's hesitation, "Papa's on the job." The third query was the most baffling. "Then were is your *father*?" Bewilderment sat on three miniature faces till there was a sudden burst of every bit of information at the eldest's command: "We haven't *got* a father! We've got Daddy and Papa and Gete, and that's *all we've got*! Gete's in the dairy barn with Mama and Mama Belle's upstairs!" With that, the three who were being inquisitioned turned tail and fled, leaving, to do as they saw fit, grown-ups who weren't above embarassing little boys who had done their best to behave in a polite and manly fashion![19]

For the overall family, the decade of the 1920s, like the 1880s, was a decade of deaths: Mama Belle in August of '21; Julie at Miles City in January of '22; Dock was buried in the space beside Sallie at the last of March of '23; and in February of '24 Jane's loneliness came to an end in the house she had bought next door to Charles Alford's house on Speedway, furnishing it to be a home, she said, for her niece, Julia (then widowed).[20] But Julie Alford Davis' life was

19. This story is still being told as part of family lore.

20. A statement made in the writer's presence.

elsewhere—she sold the place to her brother Charlie and returned to her children in North Texas and Oklahoma.

In December of '25, Zackie, ignoring his doctor's advice, concerning his own bad heart, went on his annual deer hunt in the Brush Country below San Antonio. (Ten years before, following similar advice about his brother John, too tender-hearted to pass the advice on, perhaps only half believing it, he had slipped away, much to the latter's chagrin at being left behind.) Like John, he was dangerously overweight. Before shaving their Robert E. Lee beards (the beards faded while their hair remained brown) the two had looked enough alike to be mistaken for one another so long as they kept their hats on and were not standing close together. (John had a widow's peak, causing him to comb his thick hair almost in a crew cut; Zackie was a foot the taller.) Be all that as it may, hunting had always been the love of Zackie's life (his last gift to Belle was a pair of beautifully dressed deerskin rugs he brought from the taxidermist's and laid beside her sick bed) and two weeks before Chirstmas he left Austin for Encinal (the station nearest where they would pitch camp) with his son and Belle's niece's husband, Joe Callan.[21]

On Monday morning, the 14th, he walked away from camp—and did not return. When word reached Austin on Tuesday that he was lost in the brush, friends organized a search party. At Encinal a reward was posted. A Mexican fence rider came on the body just when one of the fabulous new machines—an airplane—was under consideration to overlook the pasture. Death, it was said, had been instantaneous—actually in calling distance of his companions.

It was in a summer between these deaths that Tom Alford—hunter, trapper and guide, still unmarried—came home for the first and last time. As in the old days, he visited each relative in turn, but the hours he spent with each were few and his most memorable exploit was on Walnut, where the professional outdoorsman chose to make his bed for the night under the trees beside the creek. He arrived at Lura Payton's breakfast table speckled all over as if besprinkled with paprika—the redbugs (*chigres*—"chiggers" in Texanese) had welcomed him with delight to their home in the grass. A dip in Walnut's waters helped not a whit toward taking them off and the coating of sulphur and salty bacon grease Lura applied did little to curb their zeal as they burrowed into his flesh, while it made him sweat the more. Tomorrow, he declared, would be too long to wait for his return to the Yellowstone River where forty-

---

21. Joe Callan's wife was "Dub" Daugherty's daughter, Lillian.

degrees-below-zero weather and growling grizzlies were as nothing to contend with compared to these tropical mites in this hundred degree heat![22] Eastern sportsmen who hired his services as huntsman and guide knew him as "Tex" Alford, but he was at last and forever weaned from Texas.[23]

By 1925, with Tom in Montana, Jennie with her family in California, Baldy in New Orleans, Julia in Oklahoma, Mollie in Nocona—leaving none but Charles, the councilman, in Austin, Mary Catherine's were the most scattered of Frederic and Harriet's grandchildren. Amanda was their one remaining child, living with her daughter Eula in the Austin area called Hyde Park. Of the Payton family, Ephraim and Andrew, with their families, had gone over the "Cap" to the Panhandle and New Mexico; Dora was settled with her husband, Charles Palmer (who was a nephew of Eph's and Andrew's wives) on their dairy farm at Creedmore, thirty miles south of town; Brad and Era were truck-farming in the fertile Colorado River suburb known as Govalle.[24] John Wesley, eldest of the Payton siblings, was

---

22. Reminiscence, Pauline Payton Johnson.

23. Tom Alford died, Winnett, Montana, May 4th, 1940. His obituary was spoken by an eastern sportsman over a radio network to this effect: "Tom Alford died yesterday and so passed the finest guide in the Northwest." The writer picked up the broadcast by accident and quotes from memory. For Amanda's death and burial see fn. 17, Chapter Eleven, "Rose Cottage."

24. Bradford, rather a gentle person, though like most of Frederic and Harriet's descendants, given to temper if his family or property were threatened, would be the only one in the connection to die of violence—other than war. At dawn of Wednesday, August 12th, 1937, a hold-up man, returning to Austin from a San Antonio bank robbery evidently warned of police by some signal agreed on with two teen-age girls he had picked up and installed in a house on the town side of the Govalle Woods, fled on foot from his car and $60,000 loot, to appear at the Payton home, asking to use the telephone.

Told that the Govalle line was out of order—as it had been for quite awhile—he would have entered the house to see for himself, but was knocked from the doorstep—whereupon he fired—and Brad was dead. At least so the story is told.

Norman, in his early twenties, rushing to his father, was also shot. As he lay wounded with the gun pointing at his head, his mother, Era, flung herself on the intruder, crying, "Don't you see? He can't hurt you now!" and was herself shot twice.

Back in the house were her mother and her daughter, Mabel Julia—those two locked in a closet—and her second son, Joy, with the twin boys, Joe and Jarrell. When car keys were demanded of the youngest, Leon, Norman was able to call to him: "Do what he tells you, Leon, or he will kill you, too!"

not in good health despite Lura's care.[25]

Harriet Emily's John Lafayette and his sister, Belle Alford Swartz were farming in counties east of Austin. Frances Priscilla, George's daughter, with her husband, James Vaughn Fisher, was where she had been (near Elgin in Bastrop County) since the June day in 1881 when they honeymooned, driving to Austin accompanied by their entire bridal party in buggies.[26] Myrtle and John Cato had their homes in Montana, the former, wife of a Scottish sheep baron, Percy Williamson; the latter, of Dr. John William Butler, a veterinarian in Helena. Ethel, remarried to a career army man, Colonel E. H. Tarbutton, would settle eventually on a ranch at Kyle, just out of Austin, with a small son, Edwart Cato.

Somewhere in the shuffling and reshuffling of possessions Harriet's Bible was lost. Had she been there to keep it up to date she would have needed extra pages for the more than seventy great-grandchildren by 1925. But though their addresses ranged from the Canadian border to the Gulf of Mexico and from the Atlantic to the Pacific, they managed to keep in touch with the one early home remaining in The Field—Rose Cottage. However, by 1925, even Rose Cottage was changing face.

Leon, held hostage till the car passed from the neighborhood, was then kicked out and the car very shortly wrecked on a bridge, when the abductor fled, again on foot, to woods along the Colorado.

It was his retreat from the river the following Monday to his room in a respectable Houston home that lends the grim story its one spark of humor.

Early that morning, three portly Austin men of middle age, warned that "Brad Payton's killer may still be in the bottoms," laughed the warning off and drove down-river for a day in the wilds, taking along a dozen hamburgers lest no fish be biting.

Hardly had they found a place to park when the killer stepped out from the trees behind a sawed-off shotgun. Herding the three, despite their avoirdupois and the heat, onto the front seat, he ordered them to drive the four hours to Houston while he occupied the back, breaking his long fast.

Unaware of the foursome ahead, wiry, red-headed Jim McCoy, a deputy constable who had known Brad all his life, was also on the road, bent on tracking the killer to his lair—which was already under surveillance of Houston authorities.

Inside the city limits, the three hostages were released, the license plates smeared with mud, and the surfeited refugee proceeded to his bed, where he wakened at daylight to a nightmare far exceeding that occasioned by twelve hamburgers on an empty stomach (which some have blamed for his final action). Armed men and a fire brigade surrounded the place—tear gas bombs were falling about him. In that wild moment he turned to the one friend he could trust—the sawed-off shotgun—and blew his own head off!

Perhaps he had somehow heard (as had the writer, visiting Austin at the time) that the word was out: Don't bring him in alive—that family has suffered enough without going through the horror of a trial—Era's life was then still hanging in the balance.

On the preceding Saturday, a host of Payton kinsmen and friends had heard, over Brad's coffin, a plea for the redemption and salvation of this wayward young man, his killer, and for comfort to reach his parents in a distant state: when all was said and done, it was fear that had compounded the tragedy—his fear and everyone else's, for it was not murder or rape he had sought in Govalle, but a means of escape from the consequence of one of his own evil deeds.

25. From Isabelle Dubar's notebook.

SENATOR O.C. CATO

JOHN BACHMAN Jourdan, Miles City, Montana, 1913.

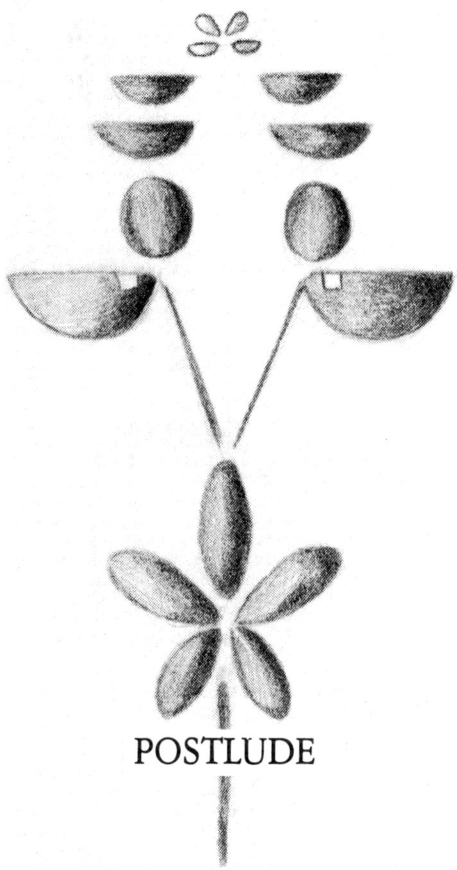

POSTLUDE

The Wheels of the Gods

Bellow of oxen and crack of wagon, whir of buggy and carriage wheel, are lost in the roar of motors; trails are buried under ribbons of concrete staccatoed with telephone poles; time appears to have speeded up—but evolution is timeless. And, though wheels in the mills of the gods grind slow (or so it was said by those older and closer to nature than we can ever be), they grind exceeding sure.

The Dream

They will walk in the garden together
Where the sunlight its shadows has thrown,
And, dreaming, pluck beautiful flowers
As they did in the days that have flown.
—Unknown (copied in Laura's Scrapbook)*

When the Cameron Road was moved to the edge of Giles land in 1925, Rose Cottage, left behind a ridge, was shut away from passing traffic. To keep up with the new flow of wheels, Eugene and Laura built with their own hands and the help of neighbors an ugly concrete house with a beautiful view overlooking the road and down Herman Kruger's pasture toward Walnut. They were immensely pleased with their handiwork—and with the growing importance of an annual community get-together called the Walnut Creek Possum Hunt, an organization that may have satisfied a desire latent since the demise of the Pickwick Club.

*Reference for this stanza has not been found. Laura may have been the composer: she wrote both prose and verse under the pen name "Olga Lewis" for local papers, particularly *The Texas State Democrat* and *The Farm and Ranch*.

The Hunt's beginning had been early in the century with evening meetings of four friends on the creek bank. Lounging one night, after watching Anthony Criswell's talented dog, Toby—half bulldog, half terrier—climb a tree to dismay of a retreating o'possum, Toby's owner, together with R. L. Jarmon, Eugene and Herman Kruger, fixed on the idea of asking Miss Claudia Boyce and the sportsman-planter, Walter L. Sanderson, to join them as self-elected-for-life directors of a Walnut Creek Possum Hunters' Association—its purpose: recognition of Travis County Pioneer Families—Jarmon to be president, Eugene, secretary.[1]

In 1925 road improvement led directly to Herman's creekside pasture as a permanent meeting place. Hunting by then was erastz with token possums smuggled in towsacks and emphasis on jubilee

1. Ledger No. 11, 1940-1941.

LAURA AS "Olga Lewis."

EUGENE as legislator.

aspects of the occasion, at one of which Jack Black, then credited with being the oldest man in Travis County, was guest speaker.

As the crowning event of the night, a tall cedar tree would be set ablaze.

Preceding the road change, Frederick Stewart Jackson, husband since 1914 to Annie Hornberger, had removed with her; their son Ocie (named for O. C. Cato); and their daughter, Dorothy Fenton (named for his mother, Fenton Sprinkle Jackson, who was Cremona's sister); to Austin, where he would be on the staff of Meyer Creamery.[2] The twenty-year-old Live Oak Spring Dairy partnership was formally dissolved August 24, 1924. Eugene's new partnership with a man of whom he knew little would end sixteen years later in unpleasantness, loss of money and a lawsuit.[3] Meantime, leaning on his popularity with the possum hunters, his membership in the Ben Hur Shrine and his pioneer status, he would be feeling his way toward the legislature.

He was scarcely seated in the capitol when Flora died (January 9, 1929). On the day of her burial, the House of Representatives adjourned its session—the one instance of her recognition as a county personage.[4]

Twice voted into office by a large majority, Eugene might have remained in politics indefinitely had he not considered it beneath his dignity to campaign indefinitely. Instead, he took the position that now people knew what he would do—if they wanted him to represent them it was up to them to see that he did.[5] He was tired of and bored with making speeches. As he saw it, entering the legislature in his fiftieth year, he had been no untried yearling or ambitious fly-by-night to use the office as a stepping stone, but a two-generation county man with the county's interests at heart. In many respects he was an anachronism; in others, ahead of his time—neither an attribute of political value except to a clever-tongued opponent who knew county interests were giving way to urban interests.

If the next returns from the polls disappointed him he did not say so. The land was there—an acreage sufficient, even without what he

---

2. *Austin American,* April 17, 1958: Interview with Mrs. Annie Hornberger Jackson.

3. Eugene's diary file details this controversy to its end in February of '41, including a complaint that the Old Place was converted into a run for greyhounds.

4. Obituary, *Austin American,* January 11, 1929.

5. Eugene in conversation with writer. It is interesting to note that his notarized campaign expenses for his first election, Place 2, 82nd District, 41st Legislature came to $166.80. Copy for his second election is missing from the papers; the third sheet—a sum commensurate with the first, was not officially filed.

had under lease, to sustain him and his two remaining sisters. The land had been hurt, he knew; and he had not been innocent of the injury—a confession he made to his diary—but with contour plowing and rotation of crops he was now doing all he could toward its healing. What he would do was retire as a country gentleman—his Aunt Jane had once said it was all he aspired to be, anyway—and devote himself as far as possible to realizing a dream he had in mind.

The dream was of a garden at the Old Place—a memorial to his mother surpassing any garden Sallie had ever dreamed of. It would be a hanging garden terracing the spring canyon and it would be included in a revival of Frederic and Harriet's homestead as an educational center for pioneer arts. To found such a center in their name would ensure their memory and that of their contemporaries as the pioneer heritage of oncoming generations in which he would have no stake—but for whom he had made a sacrifice: denying himself marriage and a family—not from lack of desire or protection of property but because there was new knowledge of heredity and there must be no risk through him of the shadow's falling on an unborn.

Instead, he would give life to the dream and it would be his child.[6]

The New Regime

Ten years after he left the legislature, on February 15, 1943, the day before his sixty-third birthday, Eugene signed a new contract with Robert Lee Helton (who was John Payton's daughter Beulah's husband), not for dairying, but for a beef cattle venture.[7] They would begin by stocking the Johnson grass pastures with "twenty-one white-face yearlings" (Herefords) branded RLH. On Tuesday, June the 25th, Laura wrote in her diary, "Calvin (Kruger) brought us 17 big roasting ears. For the first time in my life we have no corn, cotton nor small grain on our two farms. Hence the high appreciation for the roasting ears." What she wrote then was a requiem of sorts for cotton, for as surely as the sun rose and set each morning and evening on the prairies between Walnut and Gilliland, that surely, on those prairies, was the king, anointed in 1894 at the Waco cotton palace, dead.[8]

About this same time, Eugene paid his written respects both to the fallen monarch and to Sprinkle as a "ghost town" (paraphrased):

> 1900, the big year for cotton, also brought the boll weevil

---

6. Eugene in conversation with writer.
7. Ledger No. 12, 1942-1943, p. xi.
8. In Texanese, "roasting ears"—green corn—is customarily elided and softened to "ro'sen'ears."

> and the middle buster . . . highly deleterious to soil and to cotton. This was the real beginning of progressive soil erosion and lessening yield of crop. In turn, it brought to an end the long practice of farm credit and all seeking of other sources offering income than just depending on cotton, with haphazard credit; now, " 'a thing of the past.' "[9]
>
> As crop yield declined so did the population . . . From 1880 to 1940 Sprinkle rose and fell. The telephone exchange (in the Davis home) closed; the Barr store (which had passed through several unsuccessful hands after Will Jackson left it) was vacated and torn down; the railway station closed—the station foreman and crew moved away. The post office, however, was continued until 1940 when RFD (rural free delivery) was extended from Austin to serve the few remaining patrons.[10] [Not more than half a dozen voters had appeared for the last election and on August 4, 1935 the ballot box was removed to Dessau.]

Even so, no one yet recognized the incoming regime. Certainly there was no shout of "Long live the Herd!" Milking fifty or sixty cows a day, the first milking at three in the morning and the human hand the milking machine, left dairy men scant time for fanfare. And who was to suppose the smelly black stuff gushing from deep wounds along the coast—Indians knew it as Mother Earth's blood—having made its debut via the automobile, was soon to do away with the horse, the mule, the railroad, to become a despot known round the world as Black Gold?

No gold of any kind was under the Black Land: its richness had been on top—to wash away, when the sod was stripped off, down the farmer's furrows to the creeks, down the creeks to the rivers, and down the rivers to the sea.

Eugene, wheelbarrowing tons of rock for his dream terraces, when the weather and his other responsibilities allowed, wrote his Memoirs at night and on rainy days, seeking to portray what had happened in a cultural era that was past—a hundred years in which cotton came and went from the prairies between Walnut and Gilliland. This much he would have time for—this, and Sallie's hanging garden.

The center for preservation of pioneer industries, arts and crafts could wait. He would will it to the future. (See Appendix L, "Dream or Nightmare?")

---

9. Ledger No. 12, p. 141.
10. Ledger No. 9, 1937, pp. 67, 71.

L'envoi

On July 4, 1974, Frederic and Harriet's last grandchild, with their great-granddaughter, Nell, scattered flowers inside Eugene's casket, then covered it with dawn-pink blossoms symbolizing the garden built in his mother's memory. Over six feet tall, he had been a thin, handsome man his extensive family thought of as "a bit delicate." Now only the old-fashioned word "genteel" bespoke the fine-cut parchment face.

Of descendants of Edward Stanhope and Nancy Jackson Giles no bearer of the Giles name remained but Miss Annie B., a year younger than Eugene and too frail to leave her home.* Though almost as frail, among the sixty relatives and representatives of old families present was Gete's wife, Alvah—last of Frederic and Harriet's "married in" grandchildren.**

* Her deceased brother, Val C., Jr., though he married, had no child.

** Both ladies died in 1978.

Ghost Smoke

Round
the hills
the ghost smoke
rises
from the cedared gloom
with circled Indians
squatting—
pale grey ghost smoke
from the peace pipes,
rotting,
in the gloom.

Epilogue

On November 1, 1979, the *Houston Post* carried a story of the lingering demise of the city's cotton trade—a dying that began in the 30s when cotton dropped to 4 cents a pound. The *Post* expected that some member of the Cotton Exchange would list final quotations on December 31st. The story has this from Edward A. Kelly, president of the Houston Cotton Exchange and Board of Trade: Houston has not had an active cotton merchant in four or five years. The 16 story Cotton Exchange Building opened in 1924 with 2.5 million bales a year moving through the Port of Houston. Cotton made the Port of Houston; oil refineries and other industries came later. Of the cotton trade, the building remains.

INSIGNIA of the Jourdan-Bachman Park.

APPENDIX A

Annexation

Annexation of Texas to the United States was agreed upon by both republics in 1845 through a contract guaranteeing certain powers extraordinary to the State of Texas. Concluding act in the drama took place in Austin, February 19th, 1846, when Anson Jones, president of the Republic of Texas, handed over the reins of authority to J. Pinckney Henderson, governor of the State of Texas.

Major clauses in the contract are, paraphrased, (1) all public lands within Texas boundaries to be held by the state as a Public Domain—a carryover from Texas' erstwhile position as a state in the Mexican confederacy; (2) the right to divide into as many as five states, *such division to be actuated from within the commonwealth* and *not without consent of the commonwealth as a whole*—each state so formed to be received into the Union on presentation of a constitution acceptable to the United States; (3) Texas would continue responsibility for its sizable public debt.

In its first state constitution, at the insistence of Don Antonio Navarro, the one native-born member of the constitutional convention (who spoke through his interpreter), the franchise was bestowed on *all males 21 years of age except Indians not taxed, Africans and descendents of Africans.* Thus women were lumped, indirectly, with Black folk and nomadic or tribal Indians, the latter being, at the same time, distinquished from Indians "baptised and settled on the land."

Texas then saddled on the United States responsibility for "unsettled" Indians—namely Apaches and Comanches, while retaining under the governor's control its own frontier and border defensive units of Texas Rangers. To accommodate its Indian responsibility the United States established a series of military posts rainbowing above the settlements from the northeast corner of the state to the Rio Grande—meanwhile taking over the government of Santa Fé. Texas, claiming the Public Domain extended to headwaters of the Rio Grande, was ready to declare the Annexation Contract violated and was threatening secession when Washington D. C. sent down cash to the amount of $10,000,000—more than enough to cover the public debt. Texas accepted the peace offering and the boundaries were agreed upon as approximately what they are now.

Remains the Lone Star flag—no dead emblem of past glory but the symbol of vestigial remnants of a nationhood still extant in the psychology of a folk element whose legislative fiat sets it alongside the Stars and Stripes (see the *Texas Almanac* for 1976-1977) on "a flag pole of equal height," requiring only that it be made "as nearly as possible the same size." The Texas flag can never be exactly the same size as the flag of the United States because it is a tricolor—the red, white and blue bars of identical dimension with the pentagonal (five-pointed star) as the center of the blue bar which is vertical to the white and the red bars, the white bar at the top. The flag cannot be printed on any advertising material or displayed for any commercial purpose. It is flown always to left of the United States flag. It can be reversed on the staff with the red bar on top as a distress signal.

APPENDIX B

Excerpt from Diary of Edward Stanhope Giles transcribed by Eugene Victor Giles in Ledger No. 14, on April 21, 1951:

March 11 (1852)

A harder rain I have never seen fall than has fallen this morning. It has done great injury, no doubt, to growing crops—Boggy (Boggy Creek, Govalle) has washed the fence down in the field. Oh, the replanting we will have to do when the water falls. This evening Brother Thos. sent for the hands to help bury poor old Lucy who departed this life this morning.

March 12

Still raining—the Colorado higher than it has been since the year 1843, and still rising—I have no doubt but what it will do a great deal of injury to the farms on the river. About 11 o'clock—the river's falling a little but it is presumable that it will rise again as we have reason to believe it has rained a great deal toward the head of the river and that (water) has not had time to come down yet. L. P. Brown and wife—a man by the name of Harris & his family are all surrounded by water about two miles below. A good many of the neighbors have gone out to rescue them, which I hope they will succeed in doing—in after part of day—they have succeeded in getting the people out—some have lost most of their property.

Two other excerpts from the diary:

March 26

This day 24 years ago our daughter Margarite (Mrs. Wade Henry) was born.

April 22

Thos M. Giles & his family started back for Memphis when they came in Dec. 1851.

The diary closes with this statement on June 19th:

I am a good deal troubled in mind about a place to live at next year. I don't feel willing to give as high rents as I have & will have to give and the truth is I am not able to buy such a home as I want and furthermore I am not & cannot get as well reconciled to the country of Texas as I want to be.

Footnote by Eugene:

I am sure this is the same flood I've heard my mother mention as when her . . . family was living in a small cabin at Webberville. Mother said a Mr. Blair came and warned them to get out ahead of the rise but they stayed on till the rising water caught them in the night when they had to wade quite a distance thru comparatively still shallow water—said they never went back but shortly afterward moved to Grandfather's land . . . on Walnut Creek.

APPENDIX C

From *The Informer,* Houston, August 18, 1956
"The Negro's Place in the Texas Sun"

The words of the "old ones" have always been the archives of the people. This which is here transcribed is the statement of Jack Black, now deceased, but long known as the "oldest man in Travis County." His statement was taken down by Eugene V. Giles of Austin, a member of the 41st legislature, and son of a Civil War officer, who has been throughout his life deeply interested in preserving local history. The statement refers to the coming to Texas of the ancestors of many of today's well-known families.

Jack Black's Statement. "I was born in Virginia, August 10th, 1852, of slave parents belonging to Wallace Maxwell, father of J. C. and Thomas Maxwell. In 1859 Mr. Maxwell emigrated to Texas with his family—his wife and two sons—together with Sam Black and his wife, Mary Jane, and their children: Nancy, Harriet, Bill, Tommie and myself,—my brother George being left in Virginia; Mose Fraction and his wife Rody, and two children Susan and Rachel; Mack Madison and his wife, Maria, and two boys, Tom and Charlie; Fannie Allen and three children: Mary, Frank, and Lewis, and my grandmother, Fannie, whose last name I cannot remember.

"We traveled from Arlington, Washington County, Virginia to the coast by stagecoach and then by steamboat to Galveston. There the boss secured a yoke of black oxen and a wagon. Then we made the trip overland to the Custard place near what is now Sprinkle (a whistlestop on the Katy Railroad about ten miles northeast of Austin.)

"We were on the boat about a week or ten days and it took us about ten or twelve days the rest of the way. The roads—trails, really—were dry and we had no trouble with them. Most of us walked and we camped out at night. This was the early fall. Next year—1860—we made a crop of corn on the Custard place—35 or 40 bushels per acre.

"J. C. and Thomas Maxwell had come out to Texas ahead of the rest of us with five mules, and they made the arrangements for us to stay at the Custard place. Next year—1861—Mr. Maxwell bought land from Mr. Allen Burdett, who originally owned several thousand acres on Big Walnut Creek. Land was very cheap then.

"The Giles Burdett mill on Little Walnut was run by water power and ground both corn and wheat for the community. A four-year-old steer sold for ten or twelve dollars and cows for six or eight dollars.

The neighbors "swapped beef." (That is one would kill a beef and "send the meat around"—then the next week the next neighbor would kill.) We used a spinning wheel and loom. We had our own tanyard and made our own shoes, as well as our clothes. We made sorghum molasses every year and used it to sweeten coffee—when we had coffee! Mostly we used parched wheat and barley to make coffee.

"There was another mill powered by oxen belonging to Mr. William Hill in what is now the Gregg Community. Th nearest cotton gin in 1862 was on the Jim Manor Place—now Manor Town. This was a horse-powered outfit. Later (after the Civil War) the J. C. Maxwells had a mule-powered gin, V. S. Edrington was the engineer. Mrs. Maxwell fed the one-stand gin and they put out ten bales of cotton on a long day's run! Cotton seed had no cash value and were dumped in piles out of the way.

"The tools used in making a crop were the turning plow and the double shovel single-stock sweep. Corn and cotton were planted by hand. Grain was cut with a scythe and cradle. Later after the wire binder came, lots of cattle were killed from swolling (swallowing) the wire binds left in the straw from the thrasher. (Wages "after the war" were fifty cents a day and board.")

"After the war closed, my father and the other Maxwell Negroes rented land from Mr. Maxwell. In 1872 my father rented land from F. Jourdan (Mr. Giles' maternal grandfather and the father of Mrs. J. C. Maxwell). My father stayed with Mr. Jourdan for twelve years. Then he bought 41 and a half acres of Rutherford land where he and my stepmother spent the rest of their lives.

"In 1877 I was married to Pauline McFarland—a widow with four girls: Evaline, Dorothy, Ella and Jonnie. Then we had four girls: Ruth, Sarah, Hattie and Jeannette. We lived on the F. Jourdan place from 1877 to 1879. Then we moved to the Henry Neans place on Walnut Creek and made a crop. In December, 1880, we moved to this place (still in Walnut Creek Community) where we lived continuously ever since, and where we own our house and five acres of land."

APPENDIX D

Deed Book 64, p. 224

Division: Harriet's Estate - Lots 7 to 17, 63 acres each, and Lot 20 to be divided by agreement.

Lot 7 - Jennie Maxwell
Lot 8 - Z. P. Jourdan
Lot 9 - George Jourdan Heirs: James, George, Jr., Fannie (Fisher)
Lot 10 - Harriet Emily's Heirs: John Lafayette, Baldy Amanda Belle (Swartz), Alfords

Lot 11 - Julia Cato
Lot 12 - John Jourdan
Lot 13 - Amanda Payton
Lot 14 - Sallie Giles
Lot 17 Mary Catherine's Heirs: Thomas, Baldy, Jennie, Charles, Julia, Mollie Alfords
Total 567 acres

Frederick's Remaining Estate: Lots 15, 100 acres; 16, 37 acres; 18, 318 acres, 19, 63 acres; 21, 12 acres; 3, 100 acres: Total 543 acres.

PLAT BOOK 1, PAGE 31

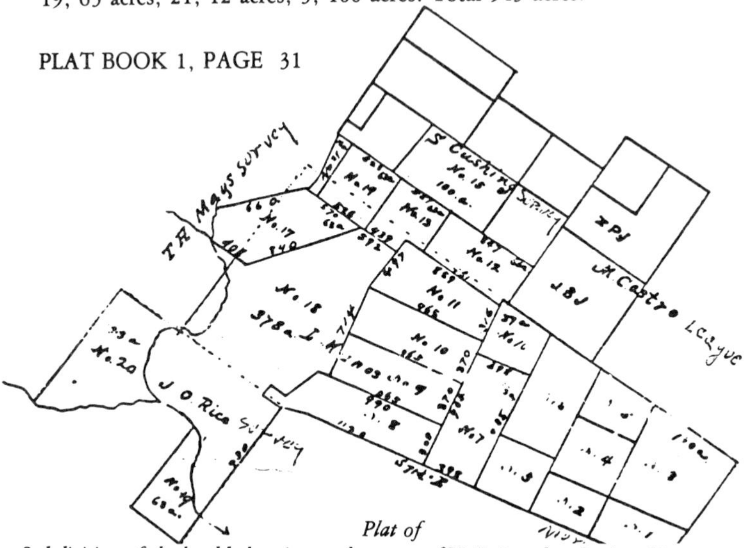

Plat of Subdivision of the land belonging to the estate of H. E. Jourdan deed and her surviving husband P. Jourdan out of M. Castro, J. O. Rice and T. H. Mays original surveys.

APPENDIX E

Lots 1 to 6 not included in division.

Austin, Texas, April 30th, 1890.

Statement for F. Jourdan Estate.

By. Sale of Part of Lot 6, Block 6	$10,000.00	
Cancellation of Insurance	116.65	
Rebate, expenses for the month of April	72.05	
To. Amount of principal, note due Brown		$6,500.00
Interest and Insurance paid by Brown		1,327.30
City Taxes year of 1889		77.35
Taxes to April 1st, 3 mos.		35.35
Paid covert & McCarty, sale of property		500.00
Alfords gross interest in the property, Retained by Fiset and Miller for T.F.A.		40.75
Amount received from sale, over all expenses		1,707.95
TOTALS	$10,188.70	$10,188.70

Different Heirs Proportions.

Z. P. Jourdan	$266.25
J. B. Jourdan	201.49
Mrs. Maxwell	201.49
Mrs. Giles	201.49
Mrs. Payton	201.49
Mrs. Cato	201.49
Alford Heirs	160.74
Belle Schwartz & Jack Alford	201.49
Mrs. Fisher	71.96
TOTAL	$1,707.

APPENDIX F

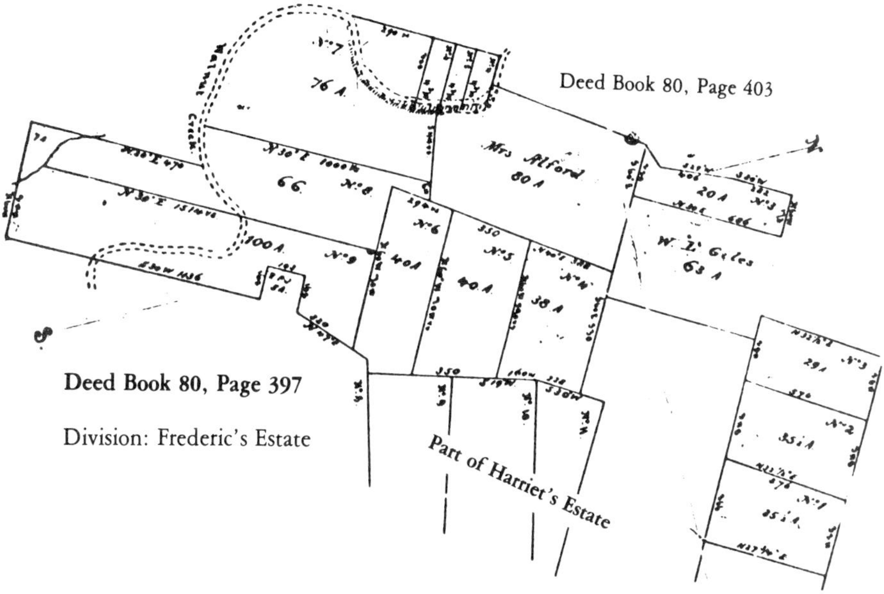

Lot 1 - Jennie Maxwell 35½ acres
Lot 2 - Amanda Payton 35½ acres
Lot 3 - Z. P. Jourdan 49 acres (2 tracts 29, 20)
Lot 4 - Six Alfords 38 acres (2 tracts 38, 4¾) Mary Catherine's heirs
Lot 5 - Two Alfords 40 acres (2 tracts 40, 4¾) Harriet Emily's heirs
Lot 6 - Three Jourdans 40 (2 tracts 40, 4¾) George Jourdan's heirs
Lot 7 - Julia Cato 76 acres
Lot 8 - Sallie Giles 66 acres
Lot 9 - John Jourdan 73 acres

W.I. GILES—63 Ac.

MRS. ALFORD—80 Ac., 76 Ac., 66 Ac., 100 Ac., 40 Ac., 40 Ac., 38 Ac., 20 Ac., 29 Ac., 35½ Ac., 35½ Ac.

Lot 4, Thomas, Baldy, Jennie, Charles, Julia, and Mollie
Lot 5, John Lafayette and Baldy Amanda Belle (Swartz)
Lot 6, James, George, and Fannie (Fisher)

APPENDIX G

The Cedar Rail Fence

Picturesque, but lost to progress.
—Condensed from Giles essays

Black land settlers penned their milk cows and work stock with cedar rails hauled down from "the Mountains" as they called the cedared hills and canyons of the Colorado.[1] Many acquired small holdings, woodlots, so to speak, around Jollyville from which to bring fencing and the fuel for their stoves and fireplaces.[2]

They built two types of fences: the "straight rail" and the "stake and rider" or "worm" fence. To construct the latter, split rails eight to ten feet long were laid horizontally in zig-zagged sections as high as desired, each rail over-lapped at the end six inches or so above or below the end of a rail in the next section. The "stakes"—unsplit rails—were then set in pairs over these lapped corners, angling at approximately 45° when tied together at the top to form Vs. Into the Vs or troughs heavy unsplit rails—"the riders"—were laid. Interstices between the rails were homes for birds, bees and butterflies.

"This was an effective fence except for certain critters smart enough to slide rails out of their way even when the fence was not wet and slick from rain. We (the Gileses) had one such expert—Old Lil—when she could not budge a rail, that fence was unbudgeable!"

A "straight fence" was constructed by setting pairs of posts in the ground, the pairs some ten feet apart, each pair with just enough room between to lay rails horizontally with the heads of the split rails over-lapping five or six inches in the post cradles. A tie-wire was wrapped round the posts near the bottom and the rails lapped above this wire to the desired height; then the posts were bound together at the top by a second tie-wire.

This construction had three distinct advantages over the worm fence: it took up much less space, required fewer rails and held the rails more firmly in place.

Cedar is a lasting material except that it is very flammable so that the only threat to a rail fence was a grass fire. Local raconteurs tell of a hot dry season when such a hazard turned toward Walnut and Mrs. Fritz Kruger was the first person out with plow and oxen to furrow down the tall prairie grasses between the community and approaching destruction. Also it is said that one neighbor, Martin Wieland, had rail-fenced an entire section of land—640 acres or one square mile.

The only cost of such a procedure was the labor involved—and at that time what price did any farmer place on his own labor or that of any member of his family?

Despite its qualities, the lifespan of the picturesque rail fence was cut to less than half a century by the advent of barbed wire in the 1880s.[3] When this cruel-to-animals but labor-saving device for men first came into use a law was passed requiring a one-by-six inch center board to be set in each fence so as to separate the strands into two groups, one below, the other above this board. The strands and the boards were attached to single upright posts set ten or twelve feet apart. Each strand was made up of two wires, twisted together and interwoven every few inches with a brief length, perhaps an inch and a quarter, of the same wire, protruding at right angles and sharpened at each end so that the barbs would rip any hide or flesh with which they came in contact—even Old Lil's.

WRITERS NOTE: The Humane Society, founded in New York in 1866, may have been responsible for this board to guard against disastrous cuts from stampedes but the Society was never really active in Texas with respect to livestock and by 1900 the law was forgotten. A horse, mule or cow, plunging into the wire, becoming entangled, struggling for release, must bleed to death unless mercifully destroyed by the owner.

So much for passing of the rail fence from the Central Texas landscape.

1. The Texas Colorado and the Brazos (Los Brazos de Dios) are the state's largest rivers.

2. Jollyville is in Williamson County. Frederic Jourdan acquired 56 acres of cedar land from the southwest corner of the James Rogers Survey, (Deed Book O, p. 59 on February 2nd, 1860). In possession of squatters for nearly a century, this land is still carried on the books of the estate of Z. P. Jourdan.

3. According to the writer's father, impact of barbed wire raised the price of Texas' unimproved land from 50 cents an acre to $2.50 almost over-night.

APPENDIX H

Excerpt from "A Ride for Life on the Oglellalah Trail-1880"
The Dallas News, December 8, 1929
Interview with J. N. Atkinson

"Our herd had a lot of cows in it. I got sick enough of being with a mixed herd before we got to Oglelallalah because every morning there were baby calves that had to be killed on the bedding ground.

"One morning I had to shoot 86 of the poor little things. Of course it was better to shoot them than to let they try to straggle along with the herd until they died—besides all the trouble we'd have had with the mothers. I think one reason we never had any trouble with wolves was because we left so many dead calves behind. But I said I'd never travel with a mixed herd again.

"When we got pretty close to old Fort Griffin we stopped and camped for a few days to give the horses some attention and make some repairs to the chuck wagon.

"One day Minelli (the trail boss) and three of us cowboys started to ride into town for supplies . . . I saw up in town a pile of buffalo hides that must have been a block long and as high as the hides could be stacked from a wagon. Apparently they had driven along each side of the pile and thrown on hides from both directions. (Asked by the writer what was to become of the hides, Mr. Atkinson said they were left there waiting for the railroad to come through.) The country round was covered with buffalo skeletons, so much so that it was dangerous to ride fast on the prairie, and I was told afterward that after the railroad came through somebody gathered up those old bones and shipped them off for fertilizer and other products. . . .

"After we passed the Indian Territory and were getting close to Dodge City, Kansas, the farmers began to meet us. They had heard that we were coming with a herd of cows and about the calves we were killing every day. They came with buffalo hides slung over the wagons to get the calves. Minelli was glad to get out of having the calves shot. The farmers would fill the wagonbeds and the buffalo hides underneath with the little fellows; they'd drive the cows alongside to feed them. After the calves were several days old they could be put out to walk and more calves loaded into the wagons.

"A fellow who went through that country next year told me one of the prettiest herds of yearlings he ever saw was the herd those farmers raised from our spring calves."

APPENDIX I

Definitions
ETHNICITY AND NATIONALITY
Seminar
IX INTERNATIONAL CONGRESS - ANTHROPOLOGICAL and ETHNOLOGICAL SCIENCES,
Chicago-August, 1973

NATIONALITY is a legal tag without which one is a man without a country—as the case of the man in the Canadian novel, *High Places* (Arthur Hailey, Doubleday, 1957) who could never leave ship because he was unable to prove he had been born anywhere.

ETHNICITY is a psychological-emotional sense of identity felt by an individual or by a community because of noticeable habit patterns or physical characteristics, or both, the whole enforced by the common socio-historical experience (mythos) or those having the same differentiation(s).

It is to be noted that this psychological-emotional sense of identity is a reaction between two or more ethnic groups - or of an individual within a stranger group - each reacting on the other in differing kind and degree but each forming an integral part of the environment of the other.

Prepared by M. Jourdan Atkinson

JIM AND JANE Maxwell at Barbecue, 1909.

MOSES FRACTION and Fannie Tucker, who was Jim Maxwell's childhood nurse.

APPENDIX J

Texas State Gazette, February 13, 1854, Vol. V, No. 25, Page 169, Col. 2
STONE'S FERRY, FAMILY GROCERY, BLACKSMITH SHOP, &C.

We would inform the public, that *Stone's Ferry,* just below the City limits, is now in first rate condition. The Boat is new and substantial, as good as can be found on the river, and a *careful, Polite* and competent ferryman will be on the ground *at all times* to manage it. There shall be no delay that human effort can avert, in crossing at this Ferry. The roads leading to it from all directions are in first rate condition—made chiefly at our own expense, and we claim this as the best and nearest road from the City of Austin to San Antonio, Seguin, Gonzales, Lavaca, &c., &c. On the western bluff of the river, we have a *Family Grocery* where every article usually obtained at such establishments can be had at the lowest cash prices of the country; and every suitable arrangement has been made for the comfortable entertainment of man and horse. Travellers can at all times obtain here supplies of provisions and provender, and at fair prices—upon this they may rely.

Connected with the premises is a *Blacksmith Shop,* where all work in that line will be done promptly, cheaply, and in the most satisfactory manner. Give us a trial. *Stone's is a licensed Ferry.*

Sam Stone & Son

Stone's Ferry, near Austin, Nov. 10, 1853.

APPENDIX K

Austin American **1909**

BARBECUE FOR HIS EX-SLAVES

UNIQUE EVENT CELEBRATED YESTERDAY AFTERNOON BY J. C. MAXWELL ON HIS FARM NEAR SPRINKLE.

AFTER FIFTY YEARS

Old Time Negroes Given A Dinner In Remembrance Of Their Coming To Texas Before The War.

RECALL ANTE-BELLUM DAYS

Twenty-five Negroes Came With Maxwell Family in 1858 and Family Now Numbers 148.

A reunion which, for its uniqueness, is probably without a parallel in the history of the south, was held on the J. C. Maxwell farm near Sprinkle yesterday afternoon. The participants in the reunion were the ex-slaves of Mr. Maxwell and their families. The reunion was the celebration of the fifty-first anniversary of the coming to Texas of Mr. Maxwell and the slaves he brought with him. It was also the birthday celebration of both Mr. and Mrs. Maxwell. Mr. Maxwell's birthday falls on November 12 and Mrs. Maxwell's on November 10. As a compromise they celebrated the occasion on November 11.

The idea was original with Mr. Maxwell, as probably no other reunion of the kind was ever held in this country. Mr. Maxwell came to Texas with his father in 1858, bringing with them twenty-five slaves. Of this number but seven of the original twenty-five were present at the reunion yesterday.

The families of these have multiplied rapidly, however, and at the reunion Uncle Mose Fraction, the senior of the ex-slaves, counted the children, grand-children and great grandchildren of the "family" and estimated there were 148 in all. Of this number about 50 were present at the reunion yesterday. The others have scattered, many of them going to Oklahoma when the territory was opened.

"I just wanted to see all the old negroes and their families together again," said Mr. Maxwell. "I haven't seen them together for a long time and may never see them together again. It will please them and they are good negroes, the best we have in the country."

His statement as to the character of the negroes is borne out by their conduct in the neighborhood. They are honest, thrifty and what the south terms "good negroes." Most of them run small farms and make a good living.

Thomas Madison, one of the original seven, has accumulated property worth probably $15,000. Others own and run farms near Sprinkle.

BARBECUE WAS AMPLE

A table was prepared and on this was spread ample barbecued steer and pig. The negroes stood around a table some 50 feet in length, the juicy pieces of meat being prepared as only meat can be by those experienced in this line of culinary.

"You folks will have to wait for the second table," announced Mr. Maxwell to the half dozen white guests who had been invited to the occasion. "This is the negroes' day and they will come first."

One of the most interesting of the old slaves was Aunt Fanny Tucker, who was Mr. Maxwell's nurse. She brought with her the old family Bible, which was owned by her grandmother. On the blank pages between the Old and New Testaments is recorded the birth of her mother in 1812. Aunt Fanny's birth is recorded there in 1828 and her sister, Jinny's, some time later. From there on the family tree has more numerous branches and is more complete. For a woman 81 years of age Aunt Fanny is remarkably preserved and she was the center of attraction, as she sat smoking her clay pipe among the younger negroes. The old mammy convulsed with laughter when some one spoke of her youthful appearance, saying she didn't have a gray hair in her head, when it was discovered she wore a wig.

There is no record of the birth of Uncle Fraction, but his age antedates that of Aunt Fanny. Uncle Moses has delusions, and sees spirits and his narration of things spiritual are subjects of much interest to his story of his attempts to dig for hidden gold in his neighborhood and how he has always been frightened away by a regiment of gaudily dressed soldiers, who always disappear in the hole where he has been digging.

The seven negroes at the barbecue who came from Virginia with the Maxwells are Fanny Tucker, Moses Fraction, Thomas Madison, William Shanks, Louis Allen, Jack Black and Amy Shanks. All are over 70 years of age, the majority of them being in their 80s. This group was photographed, as were other groups of the negroes.

Ante-bellum days were recalled by this unique gathering. The Maxwells came from Virginia in the fall of 1858. Then the war broke out early in 1861 and Mr. Maxwell was one of the first to go. The negroes stayed and managed the plantation. The barbecue recalled one time when Mr. Maxwell was away and the negroes butchered 50 hogs on the plantation and had them sent to the army. When the end came there was little left. Two thousand dollars were due on the farm and $2000 had accumulated in interest. The land was purchased from the Swenson Land company in New York and this company remitted all but the principal of the debt. This generosity was remembered kindly as it was all that allowed the family to remain in Texas.

The slaves were freed in June of 1865, but they remained with their former owners until the end of the year. For this they were guaranteed enough food and clothing to start out in their new life of freedom. The Maxwell negroes were treated well, most of them being allowed to rent small tracts of land on which they could make a living. The old negroes have all settled about Sprinkle and have done well there. They never forget their places. They are treated with kindness and give kindness in return.

Mr. Maxwell believes he has been successful in transferring the negroes from slaves into free people, a problem which puzzled all slave holders.

It has not been decided whether the barbecue will be repeated and made an annual affair, but the first reunion in celebration of an even 51 years ago was a success.

APPENDIX L

DREAM - OR NIGHTMARE?

The Cemetery

The current history of Frederic and Harriet's homestead is involved, legally, with four documents concerning their plantation cemetery—the future of which became as much an obsession with Eugene as the botanical garden to be dedicated to his mother and the preservation of the old house on the creekside. More so, perhaps, since, exclusive of the plot in the Fiskville Pioneer Cemetery where his Giles grandparents and his little brother Willie are buried, it contains all his immediate family's graves—plus nine others. When two of the documents were written and signed, his sister Laura, eleven years his senior, had already faded into the twilight zone obscuring reality that he thought of as "the family curse."

The cemetery does not lie within the seventy acres of the Jourdan-Bachman Park. It lay in the division of properties that fell to Mary Catherine Jourdan Alford—Sprinkle's heirs and the first pertinent clause pertaining to it is in the deed of C. F. Alford and wife to B. R. Payton, September 4, 1928; Deed Book 417, p. 382:

> Grantors reserve from the land here-in conveyed (181.94 acres) the grave-yard containing one half an acre of land, with the roadway thereto.

The second document involved is the Eugene and Laura Giles deed of "the old homestead" to the Heritage Society of Austin, February 24, 1956, wherein

> It is expressly stipulated
>
> (3) That the grantee herein, the Heritage Society of Austin, Inc., shall be the custodian of the cemetery near the road on the old C.F. Alford farm (at this time owned by B.R. Payton) and known as the Frederic Jourdan Family Cemetery.
>
> (5) The Botanical Garden located in the Jourdan branch on this tract is to be dedicated "To the Memory of Mrs. Sallie Jourdan Giles and the Pioneer Women of her generation."
>
> (6) If this land herein conveyed shall be abandoned by the Heritage Society of Austin, Inc., for its original purpose and shall no longer be used as a memorial park, title shall vest in the City of Austin to be used for its original purpose as a memorial park in perpetua.

There is much litigation in Texas courts today (1979) as to whose is the authority over and whose is the possession of burial grounds and cemetery plots. According to legal advice so far obtainable by the writer of this book, such authority and such possession rests with "the descendants of those buried therein"—which would throw the question open to any descendant of Frederic and Harriet Bachman Jourdan, and of Fannie Butler Thurman, minded to raise it. The sixteen graves involved are those of:

1881 Harriet Emily Bachman Jourdan
1881 Mary Catherine Jourdan Alford-Sprinkle
1883 Edwin Nicholas Jourdan
1887 Frederic Jourdan
1890 Fannie Butler Thurman
1890 Baby Ernest Thurman
1897 Sarah Ann Norman Jourdan Giles
1910 Alice Ophelia Giles
1913 William Pannell (Billie) Jourdan
1918 John Bachman Jourdan
1923 William Leroy Giles
1929 Flora Cornelia Giles

1944 Josephine Melissa Butler Jourdan
1950 Maggie Belle Giles
1961 Laura Lewis Giles
1974 Eugene Victor Giles

Of the descendants of Fannie Butler Thurman—the death of William (Buddie) Thurman (unmarried) was reported from West Texas before 1920. Neither married names nor whereabouts of his five sisters are in any family record.

The third document, Eugene's will (No. 36,325 - Volume 593, p. 819) dated July 24, 1958, reads:

> I direct that upon my death that I be buried in the Frederic Jourdan Family Cemetery located about five miles Northeast of the City of Austin, Travis County, Texas, and that a suitable head-stone be placed at the head of my grave and another of a similar nature at the grave of my sister, Laura L. Giles, upon her death.
>
> I direct that modest head-stones be placed at the gravesd of my grandparents, Edward Stanhope Giles and Nancy Jackson Giles, whose remains are buried in the Cemetery at Fiskville, Texas.

The fourth document, also dated July 24, 1958, is the Trust Indenture between Eugene, nominally his sister Laura, and the Austin National Bank. After providing for care of brother and sister during their remaining years, and the manner there-of; for specific bequests to individuals and the discharge of any debts against the estate; one finds that

> All the rest and residue . . . shall. . . be paid over to the Heritage Society of Austin, Texas, or to the City of Austin, Texas, which ever at the time of final distribution hereunder may hold title to the . . . Park . . . which includes the Frederic Jourdan Family Cemetery (a mis-statement of fact, as shown by the Alford deed to Payton) . . .
>
> That said Frederic Jourdan Family Cemetery is to be made available for purposes of last interment of those members of said families who so desire to have their remains placed therein, and said Heritage Society of Austin, Texas, or said City of Austin, is to maintain such cemetery for all time.
>
> At some place in said Frederic Jourdan Cemetery there is to be placed a small Memorial Marker showing the names of those members of said families who may be buried in other places and particularly the names of the following person: Lewis L. Giles . . . William Jourdan . . . (both died in the CSA) Willie Giles, infant brother of the donor . . .

The Memorial Park

Between the deeding of the park in 1956 and the leasing of its seventy acres March 28, 1974 (a month before Eugene's death) to THE AUSTIN NATURAL SCIENCE ASSOCIATION for 50 years, with option to renew 25 years further, the place became a night haunt of carousers and of criminals disposing evidences of stolen goods—as witness trunks of marijuana, etc., found hidden in the cedars; 10 stripped motorcycles dredged out of the ponds with a truckload of other suspect materials; plus neighboring farmers' complaints of sleep disturbed by wild goings-on in the creek bottom.

Frederic and Harriet's house (built of lumber ox teams hauled from the Bastrop sawmill, in the 1840s) broken down by a weight of baled hay, knocked down by cattle, fell into a woodpile; family members had to convince unbelieving Natural Science officials that this wreckage was once the home! The terraces and hanging gardens of the box canyon, described by an artist's daughter as a "fairyland" had vanished except for scattered stones—the picnic tables and benches Eugene made of concrete poured in frames and inset with colored potshards, broken from their pedestals and tumbled into the canyon. A bulldozer had run through the eons-old Tonkawa kitchen midden—said to be their last encampment in Travis County. Eugene never went back to view the disaster. As for the cemetery, it was overgrown by shrubbery, tall weeds, creepers (including much poison ivy-and possibly snakes) so as to make entry hazardous and identification of graves difficult.

Such was the scene when Doris Riedell, with her husband Carl, took over management in the name of the Austin Natural Science Association—and the fourteen year nightmare came to an end.

Under the new dispensation the park has become an educational center for the teaching of cultural history and the seventy acres converted to a pioneer farm. Mrs. Riedell's maternal grandmother, Roxie Tombo (Mrs. John Hickson, of Bee Caves), was a Tonkawa from whom her granddaughter learned much in the way of pioneer arts. Hundreds of children have been enrolled by their teachers for class periods on the site or by their parents for summer day camp. Youngsters brought up on concrete pavements learn the actual sources of milk and eggs, the seasonal growth of vegetables hitherto seen only on store counters, how to make soap and candles, how to barbecue meat—under Carl, a butchery professional—how to rob a beehive.

As an educational project, the area can be open to the general public only on special occasions and for specified purposes. Thus, with Riedell care of its land, its gardens, its flocks and herds of sheep, goats, chickens, ducks, geese, guineas, and what-all, the Jourdan-Bachman Park, as a Pioneer Farm, is now an asset to the City of Austin (which threatens any day to reach across the Cameron-Dessau Road and touch its boundaries with the city limits) and to the county at large.

The Estate Settlement

Eugene's estate, consisting mainly of 126 acres, was appraised after his death at $227,405.17, his short term liabilities at $59,514.57. The land found immediate sale and will presumably be held for some time by the buyer as a real estate investment. The proceeds from the sale, in accord with a proviso in the Trust Indenture, have so far been spent, for the most part, to move into the park and reconstruct appropriate buildings donated from other early estates.

Other monies (besides those from park activities) have been contributed by Junior League galas, interested individuals—and gifts have been made from school classes, such as tree plantings for parking spaces.

The cemetery, due perhaps to confusion concerning possession and authority, remains, at this writing, in pitiable condition, although the marker for those in far places and/or unknown graves has been designed and set up—as have "modest markers" for all except Mary Catherine Jourdan Alford-Sprinkle; the Giles sisters, Flora Cornelia and Maggie Belle; and—of all, most surprising, William Leroy Giles! Eugene, himself feeling touches of "the shadow" to which his father succumbed in old age, seems to have been totally unaware, when the Trust was established, that his father's grave still lacked a tombstone. Nor did the writer, entrusted with cemetery research, note its absence in the over-growth. For the same reason, graves of the two sisters were indistinguishable until small tablets (from the mortician) were kicked from the poison ivy tangle when the other stones were set. Just how Mary Catherine's stone, always listed, failed to be prepared is not known.

A restoration of the Giles plot in the Fiskville Pioneer Cemetery (off Rundberg Lane) has been ready for some time; but that cemetery, when seen in October of 1979, was sufficiently overgrown to discourage activity before being cleared by winter frosts.

What one would like to see—a possibility Eugene mentioned—is removal of the family cemetery across the present Sprinkle Cut-Off Road into the park (neither a very difficult nor a too costly endeavor, undertakers say) or else the road moved (as Natural Science wishes) and a land trade negotiated to bring the cemetery, as it is, into park extension.

DEDICATION

October 12, 1975

In the name of my cousin—Eugene Victor Giles—I hereby dedicate this Jourdan-Bachman Park—the plantation homesite of our grandparents—Frederic Jourdan and Harriet Emily Bachman Jourdan—to their memory—and to the memory of their pioneer friends and neighbors and for the use of the people of Travis County.

On behalf of our extended family, I wish to express our gratitude to the Austin Natural Science Association for their past year's work in bringing Eugene's dream to a reality—and in particular for the work of Doris and Carl Riedell.

Insomuch as Doris, through one family line, is a direct descent of the Tonkawa who were residents at this Walnut Creek Crossing for generations before Frederic and Harriet came and were—at the beginning—their nearest friends and neighbors, it is most fitting that she and Carl are now residents in charge of this park which is to be forever a homesite to come back to for us all.

M. Jourdan Atkinson

FREDERIC JOURDAN AND HARRIET BACHMAN JOURDAN
Geneaolgy

Frederic Jourdan b. North Carolina — June 2, 1811
d. Walnut Creek, Travis County, Texas — October 13, 1887
m. December 10, 1833*
Harriet Emily Bachman
b. South Carolina — October 14, 1815
d. Walnut Creek, Travis County, Texas — March 24, 1881

CHILDREN

1st	2nd	3rd	4th
Harriet Emily	George W.	Mary Catherine	William A.
5th			6th
Sarah (Sallie) Ann Norman			Amanda Bradford
7th	8th		9th
Louisa Jane (Jennie)	John Bachman		Zacharia Pannell (Z.P.)
10th			11th
Julia Elizabeth			o stillborn?

*Some family Bibles give slightly different dates. This date is from the original Bible.

1st - HARRIET EMILY

b. ____, 1834
d. July 12, 1885
m. (1) Baldy Samuel Alford, November 26, 1856, in Bastrop, Texas
b. ____, 1830
d. in CSA, November 25, 1862

CHILDREN

1st	2nd	3rd
Harriet Emily (Dude)	John Lafayette	Baldy Amanda Belle
b. August 30, 1857	b. March 25, 1860	b. June 18, 1862
d. October 8, 1878	d. May 17, 1941	d. October 28, 1958
m. Virgil S. Edrington	m. Augusta Heine	m. Oscar Swartz
November 21, 1877	February 8, 1883	November 29, 1888
(No issue)	b. March 16, 1864	b. March 3, 1857
	d. June 9, 1936	d. February 12, 1912
	(xxyyxx)	(yxyyy)

m. (2) Alfred William Waldemore, August 18, 1868
b. ____, 1844 (?), according to 1870 Travis County, Texas, census he was born in Germany—1880 Lee County (Giddings) says Switzerland.
d. ?????

2nd - GEORGE W.

b. ____, 1835
d. ____, in CSA, 1864
m. (1) Zulima C. Kuykendall, July 16, 1854, in Caldwell, Texas
b. ____?
d. ____?

Symbols indicating children of the next generation:
x - female y - male o - died in infancy, sex unknown

oo

m. (2) Lucy Payton, May 11, 1858, in Bastrop, Texas
b. ____, 1831
d. ____,

CHILDREN

1st
James W.
b. ____, 1859?
d. ____, 1888?
(unmarried)

2nd
Frances Priscilla
b. December 15, 1861
d. December 22, 1948
Elgin, Texas
m. James Vaughn Fisher
June 27, 1881
b. January 4, 1848 (Bastrop)
d. April 1, 1836
Elgin, Texas
(xyyx)

3rd
George W.
b. ____, 1862?
d. ____, 1889?
(unmarried)

Lucy Payton married E. C. Moore, December 26, 1866, in Bastrop.

3rd - MARY CATHERINE

b. ____, 1837
d. March 29, 1881
m. (1) William Thomas Alford, January 28, 1858
b. ____, 1829
d. ____, 1873
m. (2) Erasmus Frederick Sprinkle September 20, 1880 (no issue)
b. March 5, 1813
d. January 25, 1906

ALFORD CHILDREN

1st
Thomas Frederic
b. August 11, 1860
d. May 4, 1940
Winnett, Montana
(unmarried)

2nd
William
b. ____, 1862?
d. ____, 1878?
(unmarried)

3rd
Baldy Samuel
b. ____, 1864?
d. ____, 1940
m. Katherine Agnes Smith (Katie)
b. ____, 1893?
New Orleans, Louisiana
(xy)
d. December 17, 1916
New Orleans, Louisiana
m. (2) Nealie Morris
b. ____?
Austin, Texas
d. ____?
(no issue)

4th
Jennie Louise
b. June 22, 1866
d. May 7, 1963
m. Thaddeus Dale McCall
____, 1895
Nocona, Texas
b. January 30, 1867
Calvert, Texas
d. August 31, 1958

ALFORD children continued

(yyy)
5th
Charles Ferdinand
b. ____, 1868?
d. August 6, 1946
m. Vallie Billingsly, 1892
b. ____, Bastrop, Texas
d. February 20, 1952
(xyxy)

6th
Julia Harriet (Hallie)
b. August 5, 1870
d. September 23, 1963
m. Robert Thomas Davis
1892, in Nocona, Texas
b. July 26, 1868
Memphis, Tennessee
d. April 27, 1926
Durant, Oklahoma
(oxxyxxyx)

7th
Mollie D.
b. September 6, 1871
d. October 29, 1930
m. Robert Browne Clark
April 2, 1891
at home of O.C. Cato
in Nocona, Texas
b. February 5, 1867
d. August 21, 1912
Nocona, Texas
(xxyy)

4th - WILLIAM A.

b. November 14, 1840 (1st child of Frederic and Harriet to be born in Texas)
d. in CSA, March 13, 1863 (in federal prison)
(unmarried)

5th - SARAH (Sallie) ANN NORMAN

b. October 3 (?), 1842
d. March 28, 1897
m. William Leroy Giles, May 3, 1866
b. December 22, 1831
d. March 28, 1923

CHILDREN

1st
Alice Ophelia
b. March 25, 1867
d. March 17, 1910
(Easter Sunday)
(unmarried)

2nd
Laura Lewis
b. March 31, 1869
d. July 23, 1961
(unmarried)

3rd
Flora Cornelia
b. October 1, 1871
d. January 9, 1929
(unmarried)

4th
Maggie Belle
b. September 20, 1873
d. January 9, 1950
(unmarried)

5th
William Leroy
b. March 10, 1876
d. June 29, 1878
(unmarried)

6th
Eugene Victor
b. February 16, 1880
d. June 29, 1974
(unmarried)

6th - AMANDA BRADFORD

b. March 27, 1845 (at Evergreen on the Yegua, Washington County)
d. February 4, 1932, Austin, Texas
m. Rufus Daniel Payton, May 9, 1865 at Bastrop, Texas
b. ____, 1828
d. ____, 1889

CHILDREN

1st
John Wesley
b. June 11, 1866
Bastrop, Texas
d. September 30, 1929
Sprinkle, Texas
m. Lura Selma Lentz
1887 at Bastrop, Texas
b. Red Rock, Texas
d. January 9, 1958
at Austin, Texas
(xyxxx)

2nd
Ephraim Frederic
b. July 14, 1868
d. June 27, 1940
Slayton, Texas
m. Margarite Fulkes
December 18, 1889
on Walnut
b. July 24, 1870
d. April 28, 1931
at Slayton, Texas
(xyxyxyx)

3rd
Andrew Jackson
b. March 29, 1872
d. February 5, 1938
Roy, New Mexico
m. Katherine Fulkes
on Walnut
b. May 31, 1873
on Walnut
d. December 31, 1959
Hoyt, Kansas
(yyx)

4th
James W.
b. ____, 1875
d. ____, 1892
(unmarried)

5th
Eula Lee
b. October 1, 1877
d. December 28, 1964
(unmarried)

6th
Daniel
b. May 14, 1880
d. November 13, 1880

7th
Dora Phoebe
b. December 8, 1881
d. June 19, 1966
at Creedmore, Texas
m. Charles Oswell Palmer
December 31, 1912
b. September 12, 1884
d. October 24, 1964
at Creedmore, Texas
(yxy)

8th
Bradford Norman
b. June 2, 1885
d. August 12, 1937
Govalle, Texas
m. Era Pearl Thorpe
May 1, 1913
b. May 18, 1890
d. October 24, 1964
(yyxyyyo)

7th - LOUISA JANE (Jennie)

b. November 11, 1847, at Evergreen on the Yegua(?), Washington County, Texas
d. February 29, 1924, Austin, Texas
m. James Corry Maxwell, October 2, 1867 on Walnut
b. November 10, 1840, in Virginia
d. April 4, 1918, Sprinkle Texas

(no issue)

8th - JOHN BACHMAN

b. April 21, 1850, Webberville, Texas
d. July 22, 1918
m. Josephine (Jody) Melissa Butler, January 16, 1873, niece of Mirinda Adkins Sneed, ward of Judge Seaborn Sneed, St. Elmo, (Austin), Texas
b. April 10, 1856
d. January 12, 1944, Austin, Texas

CHILDREN

1st	2nd	3rd
Edwin Nicholas	John Fields	William Pannell (Billie)
b. July 27, 1874	b. February 22, 1880	b. May 21, 1883
d. April 13, 1883	d. September 22, 1948	d. August 13, 1913
	m. Dona Kimbro Boyce	(unmarried)
	August 2, 1905	
	Austin, Texas	
	b. February 3, 1881	
	d. November 13, 1929	
	(x)	

4th
Mary Butler
b. April 3, 1898
d. ____,
m. Alfred Ben Atkinson
June 8, 1926
b. October 9, 1899
d. April 7, 1931
Kerrville, Texas
(xx)

9th - ZACHARIA PANNELL

b. March 20, 1854
d. December 14, 1925, near Encinal, LaSalle County, Texas
m. Margaret Isabelle Daughtery, ____, 1885
b. October 23, 1853
d. August 6, 1921, Austin, Texas

CHILDREN
Eugene Edward (Gete)
b. April 14, 1886
d. July 25, 1951
m. Alvah Mae Horton
August 2, 1905
in Austin, Texas
b. October 31, 1883
d. ____,
(yyyxxx)

10th - JULIA ELIZABETH

b. August 31, 1858
d. January 8, 1922, Helena, Montana
m. Osceola Claiborne Cato on Walnut, October 20, 1881
b. December 4, 1858 at Waco, Texas
d. May 6, 1915 at Austin

CHILDREN
1st
Ethel Chase

b. October 26, 1882 on Walnut
d. December 7, 1952
m. A.J. (Bert) Merrill, March 18, 1910, had a son, Charles Claiborn, b. September, 1911, and was granted a divorce.
m. Col. E.H. Tarbutton (Edward Heinnerwright Tarbutton), who adopted her son, Charles Claiborn.
b. April 4, 1881
d. January 16, 1955
(Son - Edward Cato Tarbutton, San Marcos, Texas)

2nd	3rd	4th
Myrtle Jane	John Ozella	Leo Elmer
b. June 26, 1885	b. August 15, 1889	b. May 3, 1892 Miles City, Montana
d. July 18, 1952	d. January 16, 1955	d. February 22, 1907
m. Percy Williamson	m. Dr. William John Butler May 1, 1921 at Miles City, Montana	
b. December 25, 1887	b. May 31, 1881	
d. June 3, 1943	d. October 29, 1948	
(x)	(yy)	

HARRIET'S FAMILY
(from family Bibles)

John Baughman		b. October 24, 1774	Orange Co., South Carolina
		d. January 23, 1860	Walnut Creek, Travis Co., Texas
	(m. 1) Mary Catherine Barbara Beneke*, August 26, 1800		
		b. August 4, 1778	
		d. August 1, 1827	
	(m. 2) Rachel Carraway, December 29, 1829		(no issue)

CHILDREN

Maria		b. August 1, 1801	d. August 4, 1863
	(m.) John Atkinson, July 17, 1817		
Ulrich		b. Sept. 22, 1802	d. September 30, 1802
Henry		b. January 9, 1804	d. September 6, 1894
	(m.) Sarah Cargill, February 19, 1824		
Charles		b. January 11, 1806	d. July 26, 1850
	(m.) Nancy Ann Peters, September 22, 1837		
Catherine		b. February 1, 1808	d. January 7, 1828
Maryan		b. May 11, 1812	d. October 3, 1888

(m.) Shirwood Young Reams, ____,

Nancy	b. March 6, 1814	d. July 16, 1856
Harriet	b. October 16, 1815	d. March 26, 1881
(m.) Frederic Jourdan, December 10, 1833		
Elizabeth	b. July 17, 1818	d. September 11, 1901
(m.) Hiram Riley, January 25, 1838		
John Beneke	b. May 30, 1820	d. ____,
(m.) Elizabeth Ritchie, September 5, 1844		
William Gordon	b. Sept. 20, 1822	d. December 5, 1882
(m.) Caroline Green ____,		

---

* There is no certain way of spelling Beneke. It does not appear in the Bibles, but appears elsewhere as Benicker, Benecke, etc.

Mary Catherine Barbara Beneke's mother was, by all accounts, Barbara, the second daughter of Count Zimmerman who had only the two children. At his death the elder daughter inherited title and estate which fell to the 2nd daughter when her unmarried sister was thrown from a horse and died in America, but she rejected the claim, having no interest in returning to Europe. The family has no given name of the County Zimmerman nor of the elder daughter.

The Baughmans are presumed to have come from Switzerland, the Benekes, perhaps, from Alsace.

**At this point in the Baughman Bible the spelling changes to Bachman.

ATKINSONS AT BELLVILLE, TEXAS

John Atkinson	b. May 4, 1793	d. January 26, 1875
(m.1) Maria Bachman, July 17, 1817		
	b. August 1, 1801	d. August 4, 1863
(m.2) Mozelle Wilkins, April 18, 1867 (no issue) (no further data)		

CHILDREN OF JOHN AND MARIA BACHMAN ATKINSON

William	b. Sept. 16, 1818	d. Sept. 30, 1827
Sarah	b. April 18, 1820	d. November 29, 1854
(m.1) Levi Thomas Bostic, September 30, 1838		
(m.2) Reuben Plummer, April 17, 1845		
(m.3) Benjamin Cheek**, ____,		
John Pinkney	b. Nov. 20, 1823	d. January 6, 1865
(m.) Martha Ann Cheek, July 30, 1857		
	(2) J.N. Wilson	
	(3) ____ Barlow	
Catherine Virginia	b. December 9, 1825	d. February 20, 1899
(m.) William Lyons, April 20, 1854		
Mary Ann	b. December 8, 1827	d. December 23, 1827
Alexander	b. Dec. 11, 1828	d. October 4, 1863
(m.) Sarah Jane Harrison, October 19, 1858 (no issue)		
Nancy Lurany	b. August 26, 1830	October 14, 1900(?)
(unmarried)		

Martha	b. August 5, 1832	d. ____,
(m.) Charles W. Spence, October 5, 1865		
Henry Turner	b. June 9, 1834	d. July 26, 1846
Louisa Jane	b. October 14, 1836	August 20, 1894
(m.) Wyatt M. Brooks, September 27, 1855		
Rebecca Caroline	b. January 11, 1839	November 28, 1839
Solomon	b. July 14, 1840	d. April 19, 1894
(m.) Rebecca Holmes, November 11, 1877		
James	b. April 14, 1844	d. January 25, 1846

---

* Shirwood Young Reams (m.1) Mary Ann Bachman, one of Maria's younger sisters.

** Benjamin Cheek, father of Martha Ann, John Pinkney's wife.

EDWARD STANHOPE GILES AND NANCY JACKSON GILES

Edward Stanhope Giles
b. August 6, 1798, Sommer County, Tennessee
d. July 7, 1877, Travis County, Texas
m. Nancy Caroline Jackson Giles, ____, 1825, daughter of Stephen Jackson
b. April 20, 1798, Farquir County, Tennessee
d. June 1, 1887

CHILDREN

1st	2nd	3rd
Mary Marguerite	Elizabeth A.	William Leroy
b. March 21, 1827	b. ____, 1830(?)	b. December 22, 1831
d. March 9, 1913	d. February 21, 1911	at Boliver
m. Wade Henry in 1849(?)	at Cornhill (now Jarrell),	Hardeman Co., Tenn.
b. ____, 1823	Texas	d. March 28, 1923
d. ____,	m. M.R. Land	m. Sarah Ann Norman
(yyyxyxy)	November __, 1849	Jourdan
	b. ____, 1809	May 3, 1866, on Walnut
	d. ____,	b. October 4, 1842
	(xxxxyyy)	d. March 28, 1897
		(xxxxyy)

4th		5th
Lewis Lucullus	Valerium Cincinnatus	
b. September 17, 1839		b. January 26, 1842
d. December 25, 1861	d. June 30, 1915	
at Galitan, Tennessee		m. Lulu (Lou) Barnhart
in CSA		December 25, 1873
(unmarried)		b. October 30, 1850
		d. November 19, 1932
		(xy)

BIBLIOGRAPHY

Atkinson, M. Jourdan. *Creolism and Racism on the Gulf Coast.* Roots Magazine Series, No. 4, 1977 (45 pps.).

Indians of the Southwest. Naylor, San Antonio, 1963.

The Texas Indians. Naylor, San Antonio, 1935.

Barnes, Lavonia Jenkes. *The Texas Cotton Palace.* Pub. by Heritage Society of Waco, Texian Press, Waco, 1964.

Barrett, William E. *The Lilies of the Field* (fiction). Doubleday, New York, 1964.

Black, Richard W. *The Life and Diary of Richard W. Black, a History of Early Times in Uvalde* (booklet). The Catch-Up Press, El Progreso Club, Uvalde, Texas.

Cohn, David L. *The Life and Times of King Cotton.* Oxford University Press, New York, 1956.

Cox, G.W. *Pioneer Sketches* (booklet). Pub. by Montague County Historical Society, 1911. Bicentennial reprint, 1975.

Dixon, Samuel Houston and Louis Kemp. *Heroes of San Jacinto.* Anson Jones Press, Houston, 1932.

Gard, Wayne. *Fabulous Quarter Horse: Steel Dust.* Duel, Sloan and Pierce, New York, 1958. Revised edition, Lancaster (Texas) Bicentennial Committee, 1975.

Giles, Val C., *Memoirs.* See Laswell.

Inglehart, Fannie Chambers Gootch. *The Boy Captive of the Texas Mier Expedition* (fiction). J.R. Woodward, 1909. Passing Show Publishing Company, San Antonio, 1910.

Laswell, Mary. *Rags and Hope, Memoirs of Val C. Giles, 1861-65.* Coward-McCann, New York, 1961.

Les Combatants Francais de la Guerre Americaine, 1778-1783, Geneological Publishing Company, Baltimore, 1969. (Available for consultation, Clayton Geneological Library, Houston).

Mayhall, Mildred P. *Indian Wars in Texas.* Texian Press, Waco, 1965.

Malley, F.W. *Report on the Boll Weevil* (booklet). Van Boekman & Schultze, Austin, 1902.

Miller, Thomas Lloyd. *Bounty and Donation Grants of Texas, 1835-1888.* University of Texas Press, 1967.

Panorama of Nocona Trade Area. Pub. by Nocona's Bicentennial Heritage '76 Committee, S.T.J. Printing Co., St. Jo, Texas.

Price, Lucie Clift. *Travis County Marriage Records, 1840-82.* Private printing, Austin, 1973.

Ray, Worth S. *Austin Colony Pioneers.* Pemberton Press, Austin, 1949.

Roemer, Dr. Ferdinand von. *Texas.* A. Marcus, Bonn, Prussia, 1849. Tr. from the German by Oswald Mueller. Standard Printing Company, San Antonio, 1935. Reprinted, Texian Press, Waco, 1967 (Recommended by Atkinson as assigned reading for all Texas history classes above the elementary level).

Saunders, George W. *The Trail Drivers of Texas.* Compiled and edited by J. Marvin Hunter, Bandera, Texas, 1924.
Schulz, Ellen D. *Texas Wild Flowers.* Laidlaw Brothers, Chicago, 1928.
Smithwick, Noah. *The Evolution of a State.* Gammel Book Company, Austin, 1900.
Spencer, Lilian White. *Arrowheads* (Poems). Parade Publishing Company, New York, 1929.
Steen, Ralph Wayne. *The Texas News.* Compiled by Steen, Steck, Austin, 1955.
White, Gifford. *The 1840 Census of the Republic of Texas.* Pemberton Press, Austin, 1966.

Articles, Brochures, and Manuscripts

Atkinson, M. Jourdan. "Ride for Life on the Oglellalala Trail." *Dallas News,* December 8, 1928.
"Familiar Sayings of Texas Settlers." *Texas Folklore,* Vol. 5, 1926.
"Pioneer Folk Tales." *Texas Folklore,* Vol. 7, 1928.
Anderson and Clayton Company. "The Story of Cotton" (brochure). Houston, 1934.
Berlet, Sarah Groce. "Life of Jared Groce and his Eldest Son, Leonard Waller Groce." (ms.) 1936.
Bluebonnet (article on) *Houston Chronicle,* March 4, 1973.
Brown, Frank. "Annals of Travis County and the City of Austin from the Earliest Times to 1874." (mss. 15 vols.) Austin and Travis County Collection, Austin Public Library.
Carpenter, C.T., "King Cotton's Slaves." *Scribners,* October 1935. Condensed, *Readers Digest,* December 1935.
Crockett, Fannie. "Parsons Female Seminary, Manor, Texas," (ms.) Austin and Travis County Collection, Austin Public Library.
Hall, Capt. Ray. "The Horse that Proved his Worth," (Steel Dust) *The Examiner,* McKinney, Texas, December 6, 1956.
McCoskey, Betty M. "The Kuykendall Family in Texas." (ms.) Texas State Archives.
Swofford, Minnie Ray Bachman. "Our Bachman Ancestors and Our Kin." (mms. 232 pps) Procurable from the editor and compiler, Lubbock, Texas, 1979.
Walsh, Captain W.G. (b. September 23, 1836, d. August 24, 1924). Series of 15 articles beginning January 27, 1924, *Austin Statesman.* Austin and Travis County Collection, Austin Public Library.

PAST HISTORY

Kingdom come, kingdom go
Where Walnut Creek
And Gilliland flow,
Where roadrunners run
And raincrows crow
And there used to be deer
And buffalo—
Where the Tonkawa were
Till the farmers came,
Making the land never
The same again
Where Walnut Creek
And Gilliland flow,
Down river,
To the Gulf of Mexico.

M. Atkinson

INDEX

Family members running through the book are not named in this Index but will be found in The Genealogy Tables.

Cameron
Little Walnut
Walnut C
to Houston, Galveston
& The Gulf of Mexico
Gilliland Creek
Boggy Creek
Govalle
Austin
Hornsby's Bend
Hunter's Bend
Stone's Ferry
Elm Creek
Webberville
Saint Elmo
Colorado River
Sneed Mansion
Onion Creek
Bluff Springs
to Bastrop